OPTOELECTRONICS
A TEXT-LAB MANUAL

Morris Tischler

President
Science Instruments Co.
Baltimore, Maryland

McGraw-Hill Book Company

New York   Atlanta   Dallas   St.Louis   San Francisco
Auckland   Bogotá   Guatemala   Hamburg   Johannesburg   Lisbon
London   Madrid   Mexico   Montreal   New Delhi   Panama   Paris
San Juan   São Paulo   Singapore   Sydney   Tokyo   Toronto

Sponsoring Editor: Paul Berk

Editing Supervisor: Larry Goldberg

Design and Art Supervisor/Cover Designer: Meri Shardin

Production Supervisor: Laurence Charnow

Text Designer: LMD

Cover Photographer: Geoffrey Gove

Technical Studio: Accurate Art

Library of Congress Cataloging in Publication Data

Tischler, Morris
    Optoelectronics: a text-lab manual.

    1. Optoelectronics.   2. Optoelectronics — Laboratory
manuals. I. Title.
TA1750.T57   1985   621.36'7   85-10931
ISBN 0-07-064786-0

Optoelectronics: A Text-Lab Manual

1 2 3 4 5 6 7 8 9 0 SEMSEM 8 9 2 1 0 9 8 7 6 5

ISBN 0-07-064786-0

CONTENTS

I started with the crystal set. It was, and still is, one of the great marvels — music and voice received through the medium of a rock. I learned by teaching and experimenting with tubes, tube receivers, and transmitters. When my students and I were comfortable with these, we were introduced to a new form of voice reception. This time it came from sand-silicon. It was the transistor. I marveled at the crystal set, but the integrated circuit (IC) and large-scale integrated circuit (LSIC) were unbelievable. All my circuits have gone into the IC format, and designing is now a matter of selecting the proper IC.

However, it isn't over yet. Edison gave us the light bulb, and a new technology has evolved, once again from sand and rock. It is called *optoelectronics*. The mixing of light sources with optics and electronic circuits has created a whole new way of designing. More importantly, it has created a new way of living. Optoelectronics lets us check out of the supermarket faster because a machine scans the price code on the individual product. Cameras, dishwashers, and automobiles are now wired with optocouplers, and voice communications across the country are transmitted over glass and plastic fiber links.

Optotechnology is moving as rapidly as IC development once did; the material for teaching can no longer be squeezed into one book or manual. I limited the material in this text to that which would provide students with an introduction to the technology — to enable them to move forward on their own. This text is structured for a one-semester course.

Electronics, for me, has offered a lifetime of both challenge and enjoyment. The thrill of the crystal set has not diminished. Music from a rock with a cat's whisker is still a thrill, but now there is a new challenge — optoelectronics. Please join me in the exploration of these new mysteries.

Morris Tischler

ACKNOWLEDGMENTS

Acknowledgment is given to the over fifty manufacturers of optocomponents and cables who were most cooperative in supplying technical bulletins and data. Particular recognition is given to Motorola Semiconductor Products for their *Optoelectronics* device data book; to General Electric, Semiconductor Products Department, for their *Optoelectronics* volume; to Hewlett Packard for their *Optoelectronics Designer's Catalog;* to AMP Inc. for connector data information; to Mitsubishi Rayon Co. Ltd. for plastic fiber cable (ESKA) data; to GTE Lenkurt National Semiconductor Corp. for application notes and data; and to Honeywell Optoelectronics Division, Math Associates, Inc., EDN Magazine, DuPont Cable, Pirelli Cable, and Galileo Electro-Optics Corp.

Special mention is given to my favorite coworker, Maureen S. Tischler, whose encouragement and understanding were important to such an undertaking.

Acknowledgment is also given to the following members of the Science Instruments Company: to Anne Shimanovich, for her hours of typing; to Joseph Werner, for his artwork and layout; to Warren Oliver, for assistance in validating the experiments; and to other staff members, for their guidance.

The author and publisher also wish to acknowledge the assistance of William Daughaday, who evaluated most of the experiments and made many suggestions for improvement.

Finally, I acknowledge that I learned an awful lot by having gathered and read the numerous articles, data, and reference books provided to me. Once again, I discovered that the best way to learn is by teaching.

TO THE READER

SCOPE OF STUDY
This text-laboratory manual provides a comprehensive study in the use of optical electronic devices and circuits in electric power generation, data transmission, and telecommunications. A basic knowledge of semiconductors is assumed.

OBJECTIVES
Upon completion of the classroom and laboratory study of optical electronic devices and circuits, you will be able to:
1. Measure circuit voltages, currents, and waveshapes and compare them with the data provided
2. Troubleshoot circuitry which uses optoelectronic devices and replace defective parts
3. Assemble and test newly developed optotype circuits for performance
4. Develop electronic circuits that meet specific requirements
5. Use electronic laboratory test instruments to measure voltages, waveshapes, phase shifts, distortion, and photometric power in order to determine the circuit characteristics
6. Use technical data sheets relating to optodevices to determine component replacement while making equipment repairs
7. Handle sensitive components, such as light-emitting diodes (LEDs), photodiodes, and fiber-optic cabling during installation and repair without damaging parts
8. Discuss with other engineers and technicians the operation of the optoelectronic circuits and systems covered in this course
9. Cut and splice fiber-optic cables, attach special connectors, and adapt practices to similar new materials
10. Read and understand technical articles in journals and texts on optoelectronic devices, circuits, and methods
11. Write technical reports relating to circuit development, repair, or maintenance, describing the operation, malfunction, or repair performance

Electronics is fascinating because you have many opportunities to find creative solutions to technical problems. Each new and seemingly complex piece of equipment comprises a variety of basic components. In the study of optoelectronics, some recently developed devices have been added to such standard components as transistors and ICs.

This course does not use a cookbook approach, wherein you are to follow step-by-step procedures. Rather, the instructions are purposely brief, conveying only the general idea. This leaves the thinking up to you.

Each circuit is studied to determine how it functions. Indeed, the circuit may not perform properly when it is first connected. A circuit that does not work immediately requires testing, changing of component values, and so on. If you are a highly qualified technician or an engineer, you need only the circuit concept and basic information to get started. From this starting point, with the help of your test instruments, you should be able to make a functioning circuit.

Some circuits come with equations that enable you to make your own calculations and circuit changes. Some tests are required, but others may be suggested by either the instructor or you. At all times, you should be thinking to yourself, "What if I changed the bias, varied the voltage, increased the load?" The "what if" provides a great insight into circuit design.

Several circuits can be combined to form subsystems. You may want to first consider the design, assembly, and testing of such configurations and then combine the new subsystems to form larger operating units. We suggest that a brief laboratory report be prepared on each circuit studied. The format of the report should be specified by the electronics instructor.

An electronic calculator is a valuable time saver because calculations are required for circuit changes. The parts required for each circuit are shown in the circuit diagrams. Resistors are all ¼ watt (W), unless otherwise indicated. Capacitors are rated at 25 working volts of direct current.

A word of advice. When you design or redesign a circuit, think about how you would change it if the circuit did not perform as expected. Consider several alternatives. Try your circuit, try it again, and retry it until your idea is converted to a working circuit or system.

SUGGESTED LABORATORY INSTRUMENTS
Various types and qualities of electronic test instruments are available. An oscilloscope that sells for over

$2000 is better and probably more professional than one which sells for $400 to $700. Such expensive instruments can be used, but are not required for this course of study.

LABORATORY POWER SUPPLY

To perform the experiments, well-regulated power supplies are required. A well-regulated dual power supply with ± 16 volts of direct current (V dc) at 0.2 ampere (A) and + 5 V at 0.2 A is required. The 16-V sources should be adjustable, and the ripple should be less than 5 millivolts (mV) with full load.

OSCILLOSCOPE

A 5-inch (in) dual-trace triggered-sweep oscilloscope should be used. It should be equipped with the following:

Vertical sensitivity of 10 millivolts per centimeter (mV/cm) for each channel, a frequency response dc to 15 megahertz (MHz) with dc and alternating-current (ac) inputs

Horizontal sensitivity of 200 mV/cm and a frequency response of 2 to 200 kilohertz (kHz)

Sweep speed of 1 microsecond per centimeter (μs/cm) to 0.2 s/cm

Magnification of X10

Two probes, both with direct and X10 attenuation

Test leads (optional)

FUNCTION GENERATOR

This instrument should have a frequency range of 1 Hz to 2 MHz with less than 0.1 percent distortion for sine waves, triangular waves, and square waves. The stepped and variable attenuation should be 0 to 60 decibels (dB) with 10-V output into a load of 50 ohms (Ω). The duty cycle should be variable.

VOLTMETER

The digital multimeter is a most important instrument. It is suggested that a volt-ohm-milliammeter (VOM) or a field-effect transistor (FET) VOM also be available along with a digital multimeter since sometimes two measurements may be needed simultaneously. Many different types of digital multimeter are available. A minimum of 3½ digits is standard and adequate.

IMPEDANCE BRIDGE

A capacitance or RCL bridge (one or two per laboratory) will be useful for checking component values since a capacitor marked 20 microfarads (μF) may actually measure 25 μF or more (20 percent tolerance value). The instrument should measure resistance from 0.001 to 11 MΩ, capacitance from 1 picofarad (pF) to 11,000 μF, and inductance from 0.01 microhenry (μH) to 1100 H. It should be both battery- and ac-operated.

RESISTANCE-CAPACITANCE DECADE BOXES

At least one of these boxes should be available for each workstation since it is often more convenient to vary the resistance or capacitance value by turning a knob than by inserting separate components. The suggested resistance decade box is 15 Ω to 10 MΩ in two ranges. The capacitor box should have a range of 100 pF (or less) to 0.22 μF (or more). Cases should have a ground lead. An assortment of individual resistors and capacitors could also be used.

DISTORTION METER

This meter measures the distortion of amplifiers. Usually one or two distortion meters per laboratory are sufficient. The instrument should be able to measure to within 0.1 percent and have a frequency range of 20 Hz to 20 kHz, a distortion range of 0.3 to 100 percent, and an input level of 1 to 300 V.

DIGITAL FREQUENCY COUNTER

The digital frequency counter should have a frequency range of 10 Hz to 60 MHz, a sensitivity of 30 mV or less, and a six-digit readout. Two to three digital frequency counters per laboratory are sufficient.

PHOTOMETER

This instrument with a cable adapter should be able to measure from 100 microwatts (μW) to 5 milliwatts (mW) of radiated output in the visible light and infrared range. Examples of such photometers are Model S374 from Science Instruments Co., Baltimore, Maryland; or Model C from FOTEC Co., Boston, Massachusetts.

LABORATORY EVALUATIONS

A variety of circuits are presented. Some may require construction, and others may be preassembled on printed-circuit (PC) boards. The latter emphasizes testing and evaluation rather than construction. Technical information about the devices used may be found in Appendix B. For additional technical data, refer to data publications by component manufacturers. For each subject studied, complete the questions asked and, if required by the instructor, prepare a written report complete with a drawing of the circuit and parameters. For the sake of economy, the number of optocomponents used has been minimized, and no attempt was made to use similar devices from different manufacturers. Rather, emphasis was placed on studying characteristics and the proper utilization of the component.

A few basic circuits have been selected. The same components, with possibly a few additions, can be used for variations of the circuits or in the study of many other types of circuits. A set of data manuals should be available for use by the student and the instructor.

Laboratory experiments may require specific types

of measurement that require the use of test equipment. You may not be familiar with the methods or applications of this equipment; thus, you may seek instructor assistance or establish your own procedures. For example, bandwidth (BW) measurements may be required on an audio amplifier. This calls for the audio generator source to be held at constant voltage and the amplifier output voltage to be monitored with a meter while the frequency is varied from 10 to 20,000 Hz or higher. The resulting data, visually collected in octaves, should be plotted on semilogarithmic graph paper. Here is another example: The input-output (I/O) phase relationship of an amplifier can be determined most easily by connecting the input and output of the amplifier into the inputs of a dual trace oscilloscope, overlaying the traces, and measuring the displacement. A phase meter could also be used for this test.

Typical answers to review questions are provided at the end of each experiment. This data is provided so that you can use this course as a self-instructional program. You should look at the answers only after completing all the required work.

COURSE TITLE

Each laboratory assignment will require 1 to 2 hours (h). Given the additional experimentation which you are likely to perform, it is best to be safe and allocate 2 hours. If laboratory reports are required, they should be written outside normal class hours.

Classroom discussions should also be scheduled. If all units are studied, 100 clock hours should be allocated. If the course is divided into the three main areas of applications (electric power, logic, and telecommunications), the study can be arranged as an add-on to existing courses. The subject can also be used in a general semiconductor course.

MATERIALS REQUIRED

The components are standard commercial- and industrial-grade parts which are available at electronic parts outlets. These components were chosen because they are readily available, are of good quality, and are a size that can be easily handled by the experimenter. These components are given in Appendix A. The Science Instruments Company of Baltimore, Maryland, provides complete components kits, assembled panels, and a basic trainer.

COMPONENT TOLERANCES

Generally, resistors of 5 or 10 percent tolerance can be used. Capacitors may be used with 10 to 20 percent tolerance or more. In circuits where matched pairs of components are needed, check your component values and try to match them as closely as possible. Although 1 and 0.1 percent components are available, they are quite costly. Besides, it is good experience to learn when and how to find closely matching components from a low-tolerance batch since this is often necessary in fieldwork.

CONCLUSION

This course is designed to provide:
1. Broad coverage in the use of optodevices
2. A study of circuits, parameters, and subsystems through laboratory studies which do not require special preassembled parts
3. A basic system of study which provides for creativity and flexibility
4. A selection of topics which could be related to specific fields of interest
5. Coverage of standard types of circuits which could be tailored for special requirements
6. A high degree of learning with a minimum investment of time
7. Laboratory study which does not require special preassembled parts
8. A study that is not written in cookbook fashion, but rather gives you the opportunity to work out required information
9. A study utilizing the latest techniques of optical communications
10. Comprehensive study for students of electronics in technical colleges, vocational centers, government, military training centers, and industry

PART ONE

INTRODUCTION TO OPTOELECTRONICS

Optoelectronics, as a broad definition, is the integration of electronics, optics, and light to more effectively and economically control an electromechanical operation, transfer information, or make measurements. The term *light* means both visible and infrared light. Visible light can be seen by the human eye; infrared light is below the range of human perception. Optoelectronic devices include light emitters, photodetectors or sensors, transmission lines made of optical fibers, and visual displays. A variety of connectors, light-isolated couplers, and transmissive/reflective components are used in bringing the various technologies together. Electric energy can control light, or light can be made to control mechanical movements. All this can be accomplished more effectively than was possible by either technology separately.

Optoelectronic components have been proved superior to mechanical sensing and switching, they cost less, and they are smaller and lightweight. Optoelectronic components are faster, have a longer life, and are more economical. On the negative side, this technology is just emerging, and many innovations will be made. Components will become more compact, and separate components will be integrated into smaller packages. Thus some components will become obsolete. The applications of optodevices range from the space shuttle to the clothes washer. Figure A shows some types of equipment in which optodevices are used.

Optoelectronics makes extensive use of transducers of energy. In transducers, as in our eyes, light is changed to an electric current by photodetectors (photosensors). In the early days of electricity, transducers mainly changed mechanical to electric energy and vice versa. Morse code was sent over lines on land via electricity and was reproduced by mechanical breakers that moved up and down —a far cry from the pictures now transmitted of the planet Mars.

When people started to use wires and the flow of electrons, magnetic fields were also involved. At first, the magnetic fields were low in frequency. Electric power was generated at 25 to 60 cycles per second, or hertz (Hz). This frequency, when transduced to sound, is within the human hearing range. But when people moved into the wireless era, the frequency range of human hearing was left behind. In order for people to interface with this equipment, special transducers had to be designed. Some transducers provided sound, and others provided light. Interestingly enough, humans have yet to transduce smell, taste, or the sense of touch.

All energy converters have an operating wavelength which is located someplace on the electromagnetic spectrum. Figure B shows the distribution of energy on the electromagnetic spectrum. The electromagnetic spectrum is usually scaled in angstroms (Å) or micrometers (μm). Both are units of measure which can be converted to frequency. The velocity of propagation of light is equal to 300×10^6 meters per second (m/s). In this region, magnetic waves are no longer measured in cycles per second, or hertz, but in wavelengths and millionths of a wavelength. The unit of measure used is the angstrom or micrometer.

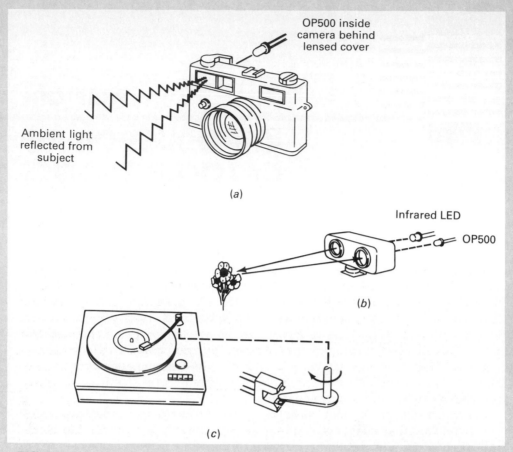

FIG. A Typical uses of optoelectronic devices. (a) Automatic light meter. (b) Automatic focus control. (c) When the tone arm on a phonograph engages the oscillating grooves on the inside of the record, an opaque object connected to the underside of the tone arm interrupts the IR light path in the interrupter, triggering the return-to-rest or record change cycle.

$$\text{Wavelength (in meters)} = \frac{300,000}{f\text{(in kilohertz)}} = \frac{300}{f\text{(in megahertz)}}$$

Wavelength, in feet, is $984/f$, in megahertz. In addition,

$$1 \text{ Å} = 3.937 \times 10^{-9} \text{ in} = 1 \times 10^{-10} \text{ m} = 1 \times 10^{-4} \text{ } \mu m$$

Also

$$1 \text{ } \mu m = 3.937 \times 10^{-5} \text{ in} = 1 \times 10^{-6} \text{ m}$$

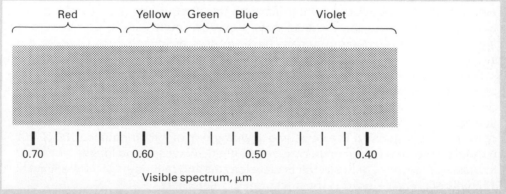

FIG. B Electromagnetic spectrum.

This is equal to 1×10^4 Å. As an example,

$$100 \; \mu m = 1,000,000 \text{ or } 10^6 \text{ Å} \qquad 1 \; \mu m = 10,000 \text{ Å}$$

As another example, a purple light of 0.40 μm is equal to 4000 Å. Particular attention should be paid to the light frequency range since optoelectronic devices work in regions of visible and infrared light. Many manufacturers produce only infrared emitters and sensors. Light-emitting diodes (LEDs) which radiate in the infrared region (below visible light) are called *infrared-emitting devices* (IREDs).

Although optoelectronics is a new technology, optical transducers have been used as controllers for some years. Vacuum-type photocells were used to control solenoids and motors for moving doors and regulating the flow of people. The motion picture projector incorporated a light transducer along with mechanical devices to create sound from photographically produced sound tracks. Today a similar technique is used to read codes on products at the supermarket check-out. The newer devices are more versatile, and a greater variety are available for circuit design.

Advances in semiconductors have contributed enormously to the growth of optoelectronics. The invention and development of semiconductor devices, such as the PN junction and transistors, gave real impetus to the emergence of the optical electronics (optoelectronics) industry. In place of a tungsten lamp, heated to 2000 or 3000 K to produce light, black sand is refined into silicon disks to which are added "doped" materials such as arsenide. Under proper conditions, the junction of P- and N-type semiconductors is able to radiate light. Although this light is not as bright as that from light bulbs, we can still use the LED and add highly sensitive detectors to compensate for the difference. The LED, depending on its structure and doping, will give off light from near ultraviolet to near infrared. Units are made in plastic or metal cases, with and without plastic lenses for focusing. The LED is the most popularly used emitter, although tungsten, neon, lasers, and other light sources have been, and are, used.

Light detectors or sensors, also made from semiconductors, are used to convert light energy (photons) back to electricity. The PN junction diode, when provided with a forward current (bias), will radiate light; but when back-biased (or reverse-biased), the diode produces electron flow from light. The three-terminal transistor (NPN) can also be used as a diode amplifier. The base connection is left open, and radiant energy on the base-collector junction produces an electron flow which is increased by the forward gain (h_{fe}) of the transistor.

The LED and sensor can be used for counting. The beam of light between them is interrupted (transmissive control), and each break is recorded. The light can also be reflected off a turning wheel or object to determine its speed. The key elements are the emitter and detector, coupled so as to perform a specific function. These devices are called *couplers.* Some couplers contain the emitter and sensor, each electrically separated from the other. High-voltage isolation and low capacitance enable the device to control high-power circuits with low-voltage control. In the hospital, for example, where some patients have been electrocuted while undergoing diagnostic tests, optical couplers now separate the patient from the electric instrument.

The emitter and sensor need not be in close proximity. In fact, the sensor could be thousands of miles away. The emitted light is coupled into a clear, flexible, glass or plastic fiber, located inside a protective covering. This fiber link enables light to transverse it by bouncing back and forth off its walls. In this way, lightweight cables, carrying thousands of voices and computer data, replace heavy, costly, copper cables.

At the ends of the fiber lines might be telephone users, TV stations, and computers. In some cases, the transferred data is displayed on LED seven-segment alphanumeric displays. Light displays are an additional use of semiconductor illuminators.

The optoelectronic system is composed of an emitter, a sensor, coupler (Isolator) fiber links, and a variety of fittings. Some systems may require only a coupler, while others may use an emitter and sensor in a reflective mode.

As time passes, systems engineers and circuit designers will find greater applications for this new technology. The layperson, however, accepts the change with apathy. Calculators, computers, digital watches that play songs, highway cruise controls, pushbutton telephones, and TV sets with 1-in screens are now considered routine. But what brought it all about was science fiction which came true. Optoelectronics is not just another step; it is a leap into newer and better ways of controlling machines and communicating around the world.

SUMMARY

The generation of electricity evolved into the field of electronics, and vacuum-tube technology gave way to semiconductor technology. People found ways of making more efficient electronics components out of treated silicon, wafers, or chips. The new technology of optoelectronics includes transducers (emitters and detectors), fiber links, displays, and a variety of fittings to link into computers, telephones, and televisions.

AN OPTICAL COMMUNICATIONS SYSTEM

SCOPE OF STUDY

This laboratory experiment is intended for a student having little or no knowledge of fiber-optic communications systems. An overall concept is developed.

OBJECTIVES

Upon completion of this experiment, construction of circuitry, testing, and evaluation of data, you will be able to:

1. Describe, in general, how communications take place by means of a fiber-optic link
2. Name and describe the key components used in a fiber-optic link
3. Adjust the optical transmitter and receiver to achieve good communications

BACKGROUND

Telecommunications can take place between distant transmitters and receivers by four means:

1. Wire is the communications medium.
2. Radio waves are sent through space; this is a wireless process.
3. Microwaves are used for high-frequency communications.
4. Light, in the form of ultrahigh-frequency (UHF) magnetic waves, is transmitted through glass or plastic cables.

At the transmitter site the intelligence to be transmitted modulates a carrier wave, and at the receiver the intelligence is separated from the carrier.

In the wireless system, antennas are used for the transmitting and receiving of the signal. In optoelectronics, the linkage is made with a glass or plastic fiber-optic cable. The carrier is a light wave; instead of transmitting through a radio-frequency (RF) amplifier, a LED, an IRED, or a laser is used to transmit the signal. The light is received by a photodetector which may be a light-sensitive diode, transistor, or Darlington transistor circuit.

Plastic and glass fiber-optic cables have long since passed the experimental stage in applications to telecommunications. Telephone companies nationally and internationally are installing thousands of miles of fiber-optic cable for the conduction of voice signals. Frequency division and pulse-code-modulated signals are being transmitted over glass fibers, whose diameter is less than 10 μm.

Basically, a fiber-optic system has three major components: a light-emitting source (such as a laser or high-powered LED), a fiber-optic cable having a low level of attenuation, and a photosensor. The photosensor is followed by a variety of amplifiers and decoders.

Much of the light in communications over fiber-optic cables operates in the infrared region of 820 to 1600 nanometers (nm). Silica material, used in the making of light emitters and sensors, operates more effectively in the infrared region. So it is easy to fabricate infrared devices by using standard technology. Where a high level of energy is required, solid-state lasers are incorporated. The optical fibers come in three basic types:

1. The step-index fiber
2. The graded-index fiber
3. The single-mode, step-index fiber

The single-mode fiber can also be made in the graded-index format. Within the step-index fiber, the light bounces off the walls of the fiber, thus making a zigzag path through the fiber cable. Some light enters the plastic coating, referred to as *cladding*. The light entering the cladding may continue for a short time, but it is rapidly attenuated. The step-index fiber is often used in the manufacture of plastic fibers, where short runs are encountered. For long runs, such as in telephone communications, high-quality glass fibers, which use single-mode conduction techniques, are in general use.

In the graded-index fiber, a lensing effect causes the beam to be refocused as it travels down the fiber's length. The losses in this type of fiber are not nearly as great as in the step-index, multimode fiber. Once again, in the step-index fiber, the light zigzags through the cable, and a high level of loss occurs.

As the cable is made smaller in diameter, it becomes more effective in conducting a single-mode light wave. Although there are losses in the single-mode fiber, they are minor compared with those in the larger cables made of the step-index form. Attenuation levels of 1 to 5 decibels per kilometer (dB/km) have been experienced in glass fibers as compared to 500 to 9000 dB/km in plastic cables. Once again, plastic cables are best used for short runs, where amplifiers can make up for the signal losses.

The glass and fiber cables offer many advantages not available in metallic conductors. Some of these advantages are as follows:

1. The spacing of repeaters ranges from 20 to 70 miles (mi), compared with 1 mi for copper cables.
2. Fiber cable is immune to water and moisture conditions.
3. The light weight and lower cost of the cable make it advantageous to install.
4. A single-fiber cable can conduct as many voice channels as can be handled by a metallic cable containing 900 wire pairs.
5. The light weight of the cable suggests applications in aircraft and other mobile vehicles.
6. Fibers made of glass use silica as the base material, compared with copper and lead used in making metallic cables.

In the transmission of voice or data communications, it is extremely important that the light-emitting source and the photosensing source have high-speed characteristics. Slow rise and fall times could result in the overlapping of pulses.

Since the fiber-optic cables are conducting light, they offer a very wide-frequency response as compared with RF carrier signals used on metallic cables. In view of the wide bandpass, a much larger number of voice channels can be accommodated.

One of the problems currently encountered in using fiber cables is the variety of couplings, plugs, and jacks available for interfacing fiber cables to hardware. Standardization has yet to be achieved, although the telephone industry is more likely to standardize sooner than other telecommunication organizations.

Fiber cables, although small and fragile, are often bundled into tubes which contain a polyvinyl chloride (PVC) outer jacket. The center of the cable consists of steel fibers which give strength to the overall cable. Single-mode fibers used in such cables can handle TV channels as well as thousands of individual voice channels.

Although reference has been made to the losses which exist in fiber-optic cable, still other losses must be taken into consideration. Losses exist at the connectors where a LED or laser couples into the fiber cable. The spacing between the laser and the cable is an area of potential loss. The same problem exists where the fiber cable couples into the photodetector. The frequency responses of the cable, emitter, and sensor determine the overall frequency response of the system. Besides losses in the cable, there are losses at the photodiode sensor.

In general, the photodiode is back-biased with a very small amount of current flowing through the diode. The radiant flux causes the photodiode to change its current flow. This current flow passes through a resistor across which a signal voltage is developed.

Plastic cables range in diameter from 50 μm to over 1 millimeter (mm). In the laboratory experiment, Crofon cable is used. These cables are similar to the ESKA cable, which is a trademark of the Mitsubishi Rayon Company, Inc., of Japan. The Crofon cable, formerly made by the Dupont Company, is now being produced by the Japanese firm. This cable, made of plastic, is designed for use in the visible light range. There the cable has an attenuation of approximately 500 dB/km. When the cable is used, however, in the infrared region (above 820 nm), the cable appears as a cement wall, and light will not pass through it. Once again, for short runs the plastic cable can be used. However, it must be used in the visible light, or near-infrared, range. The cable exhibits a loss of approximately 1 dB/m (type EH4001) or 0.9 dB/m (type OE-1040).

In summary, a voice-, video-, or data-encoded signal can be transmitted through a fiber-optic cable. The carrier is a light wave, whose frequency can be in the visible light, near-infrared, or deep-infrared range. The fiber cables, using solid-state light emitters and sensors, offer many advantages over standard metallic cable. Fiber-optic communications, now in its infancy, will become widespread in the coming years.

In this experiment, light-sensitive components are used to demonstrate the principles of a basic fiber-optic link.

TESTS AND MEASUREMENTS

MATERIALS REQUIRED FOR EXPERIMENT

Active Components
SE4352 IRED emitter/connector
SD3443-2 IRED detector/connector
TL081 Operational amplifier (op amp)

Resistors	**Switches**
330 Ω	4PST Dip switch (2)
1 kΩ (2)	
10 kΩ (2)	**Capacitor**
8.2 kΩ	47 μF
100 kΩ	

Miscellaneous
Fiber cable, OE-1040, 50 cm, with connector
  plugs

The purpose of these laboratory experiments is to familiarize you with the basic concepts and some applications of light emitters, sensors, and fiber-optic cables and connectors. You will modulate an IRED, transmit a signal over a fiber-optic link, and then view the received signal coming from the photodetector.

Refer to Appendixes A and B for technical data.

1. Figure 1-1 shows the circuit diagram of the basic operating system. Check your power supplies to ensure that voltages are properly set before you connect them to the experimental circuit.

=== **WARNING** ===

AS A MATTER OF GENERAL PRACTICE, YOU SHOULD *NOT* LOOK INTO THE SENSOR END OF THE FIBER LINK OR INTO THE EMITTER WHILE THE POWER IS ON. THE RADIATION, AT HIGH POWER LEVELS, MAY BE INJURIOUS TO HUMAN EYES.

2. The system is arranged so as to transmit an alternating-current signal on an infrared beam. The IRED is forward-biased by the 330-Ω resistor connected to $+10$ V. The signal, which is ac-coupled to the IRED, modulates the dc level of the emitter. Connect one channel of your oscilloscope (vertical set at 10 mV/cm) to the output V_o (TP_3) and the second channel to TP_1. Set your function generator to a 1-kHz square wave, 0 V. Connect a 30-cm fiber cable between the emitter and the sensor.

3. Turn on your power supplies. Close switches 1d, 2d, 3d, and 4d. All other switches must be open. Now slowly increase the generator's output so that the waveshape at V_o is approximately 1 V peak to peak (p-p).

4. How much modulating signal p-p voltage V_{sig} is required at TP_1?

5. Without an input signal, how much dc voltage is on the IRED?

6. Compute the dc value of the IRED, I_f.

7. From the technical data about the emitter, determine the dc power output P_o from the value of I_f computed in Step 6.

8. Will the system pass a sine wave signal? This completes your measurements. Your instructor may suggest making other measurements such as the frequency response or pulse modulation of the IRED without the use of a dc square wave.

9. Turn off the power. Disconnect the fiber-optic cable from the sensor, and observe the size of the center fiber. The center core is 1 mm in diameter.

10. The fiber cable has the least amount of loss in the visible light range. In what range does the emitter being used produce its main light?

REVIEW QUESTIONS

On a separate piece of paper, complete the following statements with an appropriate word or words.

1. IRED emitters operate in the _____ wavelength.
2. The visual spectrum ranges from approximately _____ to _____ μm.
3. A transducer is used to _____ energy to another form.
4. The basic material in semiconductors is _____ .
5. A transmissive controller has the LED on one side of the object and the detector on the _____ side.
6. In a fiber-optic system, the two end subsystems are _____ isolated.
7. Microwaves and radar operate _____ the visible light portion of the spectrum.
8. In the phototransistor detector, the input diode is _____ -biased.
9. The core of a fiber-optic cable is made of _____ or _____ .

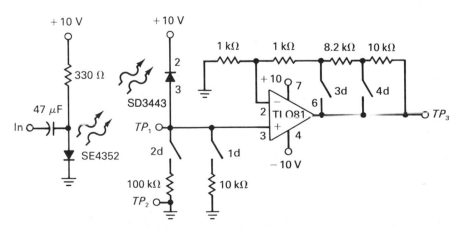

FIG. 1-1   **Fiber-optic link.**

10. In the basic communications system used, the IRED was the _____, and the detector was the _____. The cable was the transmission line.

ANSWERS TO REVIEW QUESTIONS

1. Near-infrared
2. 0.45 to 0.7
3. Convert (or change)
4. Silicon
5. Opposite
6. Electrically
7. Below
8. Back (or reverse)
9. Glass, plastic
10. Transmitter, receiver

PART TWO

OPTOCOMPONENTS

Optocomponents fall into two general categories — light emitters and light sensors. Light emitters and sensors can be further divided into devices which operate in the visible light range and those which operate in the infrared region.

A further differentiation in the various devices involves their physical structure. Different size holders have been designed for different devices. Since the industry is still quite young, light emitter sockets have not been standardized by the various manufacturers.

Light sensors are divided according to their speed of operation, frequency of operation, and ability to provide amplification. A further division relates to whether the light emitter and sensor are an integral part of one holder, such as in the optocoupler or optoisolator.

The specific application will determine whether a photodiode, phototransistor, photo-Darlington, or Schmitt trigger device is necessary. The greater the amplification of the device, the slower is its speed. In digital applications, for example, high-speed devices are generally required. If a photo-Darlington sensor is used in a high-speed data link, there is great likelihood that a trailing edge will appear in the pulses being transmitted.

Optoisolators and optocouplers are finding wide application in power control devices, where they are used in place of relays. Since the circuit to be controlled is isolated from the power source, the selection of light emitter and light sensor will depend on whether the devices are directly coupling one into the other or whether they are being interfaced by a fiber-optic cable. Should a fiber cable be used, the wavelength of light transmission for the specific cable must be taken into consideration.

The experiments in Part 2 relate to testing and evaluating various circuits which use a broad range of light emitters and light sensors.

EVALUATING LIGHT EMITTERS

SCOPE OF STUDY

This experiment provides an overview of how PN junction devices emit light and how they compare with tungsten lamps.

OBJECTIVES

Upon completion of this experiment, construction of circuitry, testing, and evaluation of data, you will be able to:

1. Describe the differences in use of an incandescent lamp and a LED light source
2. Explain in general terms how a PN junction produces light and its wavelength
3. Describe the construction of a PN junction as an exciter

BACKGROUND

The light that is observed by humans occupies a very small portion of the electromagnetic spectrum. Moreover, light has many colors which are related to the heat of the energy being radiated. Blue light is transmitted in the upper wavelengths while hues of red are seen in the lower-frequency range. Figure 2-1 shows the distribution of light from a tungsten lamp as well as the visible light spectrum seen by the eye.

As noted earlier, the higher the temperature of the lamp, the greater the visible light. The light output can be expressed in either photometric or radiometric terms. The lamp is used to illuminate a surface. In more scientific terms, the *luminous flux density* (measured in lumens per unit of surface area) is brightening the surface with incident rays of light. In radiometric terms, the energy source is radiating the surface (in watts per unit area) at a certain wavelength. Light (photometric), as perceived by the eye, falls in the wavelength range of 0.38 to 0.78 μm. One *micrometer* (radiometric term) is one-millionth (10^{-6}) of a meter. Another term for expressing the wavelength is the angstrom (Å). One *angstrom* equals 10^{-10} m; hence 0.38 μm is equal to 3800 Å. The wavelength may also be stated in terms of nanometers (1 nm = 10^9 m). A wavelength of 0.38 μm is equal to 380 nm.

The spectral radiation of light from a tungsten lamp is not equally distributed in all directions. Most of the lamp radiates in the horizontal plane (it is assumed that the lamp is set with the base down or up). (See Fig. 2-2.) The light distribution pattern depends on the physical

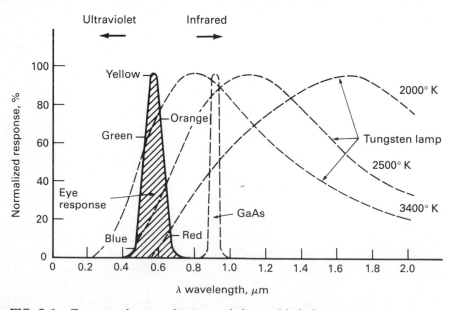

FIG. 2-1   Tungsten lamp radiation and the visible light spectrum.

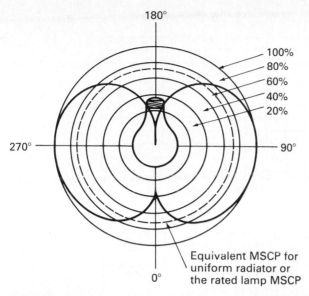

FIG. 2-2 **Typical radiation pattern of a tungsten lamp and LED.**

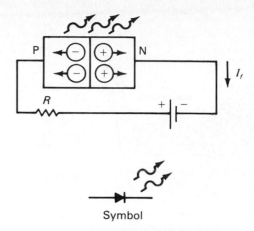

FIG. 2-3 **Forward-biased PN junction (LED).**

features of the lamp, such as filament shape, size, glass coating, and the relationship of the filament to the base. The lamp is rated in terms of its mean spherical candlepower (MSCP). The area within the dotted circle in Fig. 2-2 is the same as that of the two horizontal spheres of the lamp's polar plot. With MSCP as a unit, the horizontal radiation of the lamp equals 1.33 MSCP while the vertical output (opposite the base) is 0.48 times the rated MSCP.

The frequency response of a lamp has only a small portion of its radiation falling within the visible light spectrum. Its luminous output depends on its position with respect to the receiver (an eye or electronics detector). The lamp radiates most of its energy in the infrared region. This area is not seen, but its presence is manifested in the heat it radiates. A fireplace, as an example, has its maximum radiation in the infrared region, what we call heat.

The tungsten lamp, neon, fluorescent, and xenon tubes all work within the visible light range and are vacuum- or gas-filled. With the introduction of solid-state devices, such as the diode and transistor, the basis for a new light source became available. The PN junction diode, when forward-biased, will radiate light and other energies. Junction luminescence takes place when a low forward-biased voltage is applied to a properly doped PN junction. Depending on the structure of the junction, the LED can be made to radiate from the near-ultraviolet to the near-infrared region. Diodes which operate in the infrared region are termed IREDs. Many light (meaning visible and infrared) sources used in optical transmission systems operate in the infrared region.

Figure 2-3 shows a diagram of a forward-biased light-emitting PN junction. With forward bias, current

flow in the junction causes the holes to flow through the N material, while the electrons flow toward the P material. The electrons flow through the outer circuit and reenter the N material.

The recombining of the minority carriers (electrons) in the junction causes radiant energy to be released. This energy is proportional to the energy gap. The higher the energy gap, the higher the frequency of the emitted light. LEDs having small-band gaps made with gallium arsenide (GaAs) will radiate in the red or infrared region. Figure 2-4 shows the spectral response of an infrared light source. The angular radiation pattern ($\pm 10°$) of an emitter, with a lens, is shown in wavelengths longer than 500 nm. The LED or IRED has a highly directional output whose radiance depends on the level of diode current flow.

LEDs are rated in terms of their operating voltage V_f, reverse leakage current I_R, reverse breakdown voltage $V_{(BR)R}$, power output P_o (in milliwatts or microwatts) at a specific current I_f, peak emission wavelength λ, and spectral half-width lambda. Figure 2-4 shows the electrical and optical characteristics of a Motorola infrared-emitting diode. Although some manufacturers may rate their light output in radiometric terms, others may use photometric terms, such as luminous intensity. The *lumen* is a photometric term which is expressed in units called candelas (cd) or in smaller units, millicandelas (mcd). In either case, the transmitted output is an optical flux T, which is one of the important elements in a light communications system.

Many manufacturers rate their output in terms of microwatts or milliwatts per unit area. The area selected is the steradian (sr).

In a flat plane of a given circle, the arc around the circumference is divided into radians. When the area described is on the surface of a sphere, the angle of the cone formed from the center of the sphere to the edges of the circle is described in steradians. The steradian is equivalent to the square of the radius.

FIGURE 1 — RELATIVE SPECTRAL OUTPUT

CONVEX LENS

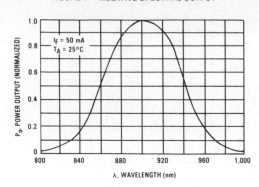

ELECTRICAL CHARACTERISTICS ($T_A = 25°C$ unless otherwise noted)

Characteristic	Fig. No.	Symbol	Min	Typ	Max	Unit
Reverse Leakage Current ($V_R = 3.0$ V)	—	I_R	—	2.0	—	nA
Reverse Breakdown Voltage ($I_R = 100 \mu A$)	—	$V_{(BR)R}$	6.0	20	—	Volts
Forward Voltage ($I_F = 50$ mA)	2	V_F	—	1.25	1.5	Volts
Total Capacitance ($V_R = 0$ V, f = 1.0 MHz)	—	C_T	—	150	—	pF

OPTICAL CHARACTERISTICS ($T_A = 25°C$ unless otherwise noted)

Characteristic	Fig. No.	Symbol	Min	Typ	Max	Unit
Total Power Output (Note 1) ($I_F = 100$ mA)	3, 4	P_O	200	650	—	μW
Radiant Intensity (Note 2) ($I_F = 100$ mA)	—	I_O	—	1.5	—	mW/steradian
Peak Emission Wavelength	1	λP	—	900	—	nm
Spectral Line Half Width	1	$\Delta\lambda$	—	40	—	nm

NOTE:
1. Power Output, P_O, is the total power radiated by the device into a solid angle of 2π steradians. It is measured by directing all radiation leaving the device, within this solid angle, onto a calibrated silicon solar cell.
2. Irradiance from a Light Emitting Diode (LED) can be calculated by

$$H = \frac{I_O}{d^2}$$ where H is irradiance in mW/cm^2, I_O is radiant intensity in mW/steradian; d^2 is distance from LED to the detector in cm.

FIGURE 2 — FORWARD CHARACTERISTICS

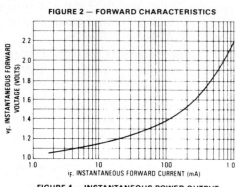

FIGURE 3 — POWER OUTPUT versus JUNCTION TEMPERATURE

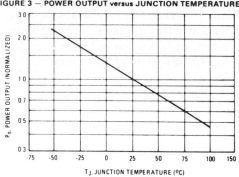

FIGURE 4 — INSTANTANEOUS POWER OUTPUT versus FORWARD CURRENT

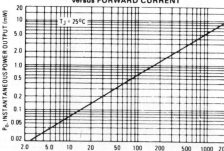

FIGURE 5 — SPATIAL RADIATION PATTERN

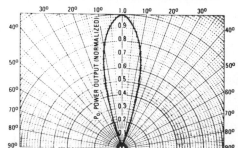

FIG. 2-4   Infrared diode, Motorola MLED930.

The measurement described in the data sheets is taken on the mechanical axis and assumes that the LED is a spherical radiator. For small angles, the rating in watts/sr describes the amount of power striking a surface at a distance from the lens of the LED. The Motorola MLED930 is a LED mounted behind a lens whose spatial pattern of radiation lies well within $\pm 15°$. If a detector is placed well within the lighted area, the detector will receive the required radiated energy.

For convenience, in systems design, the output of a LED can be stated in terms of decibels above 1 milliwatt (abbreviated dBm), thus enabling all calculations of gain and losses in a system to be summed algebraically. The LED output, flux in microwatts, is referenced to 1 mW (1000 μW). The output of the LED can be converted to the dBm unit by the equation

$$\text{Output flux } \phi_T \text{ (dBm)} = 10 \log \frac{\phi_T}{\phi_R}$$

where ϕ_T and ϕ_R are in microwatts. From the data in Fig. 2-4, the power output equals 650 μW (typical). Thus flux (dBm) = $10 \log (P_2/P_1)$, where P_2/P_1 is the power ratio. (Since P_2/P_1 is less than 1, invert and the answer denotes a loss.) The output flux is -1.87 dBm.

The output optical flux is used later in determining the losses in transmission through a fiber-optic cable to a photodetector. The overall calculation of system losses is referred to as *flux budgeting*.

SUMMARY

The incandescent lamp and other visual indicators radiate in the visible light and infrared regions. Much of the energy is radiated in the form of heat. LED semiconductors have a PN junction, which can be made to radiate at various wavelengths. The base material, type of doping, and size of the energy band at the junction determine the type of light. Many LEDs are made to limit their radiation to the infrared region. These devices are referred to as IREDs.

The radiant energy (output flux) can be focused by a lens in order to narrow its beam, hence concentrate its power.

=============== WARNING ===============

SINCE THE LIGHT IN INFRARED EMITTERS MAY NOT BE VISIBLE, YOU MIGHT THINK THAT THE IRED IS NOT GLOWING AND SO BE TEMPTED TO LOOK INTO THE LENS. CONCENTRATED INFRARED LIGHT SOURCES MAY BE HARMFUL TO THE EYE, AND DIRECT OR REFLECTED VIEWING OF THE LENS SHOULD BE AVOIDED. A SENSITIVE FLUORESCENT CARD IS AVAILABLE AND SHOULD BE USED TO SHOW THAT THERE IS AN IRED OUTPUT.

TESTS AND MEASUREMENTS

MATERIALS REQUIRED FOR EXPERIMENT

Active Devices
LED, visible light
SE4352 IRED emitter
MFOD300 Phototransistor

Resistors (5 percent)
1 kΩ (2)

This experiment relates to the evaluation of LEDs and IREDs. The diodes are forward-biased (ON), and a photo-Darlington connected transistor is used to evaluate the LED radiation.

Because the photodetector is sensitive to visible light, its input must be shielded from ambient light by the use of black tubing at least 1 in (2.5 cm) in length. Radiant light will enter through the tube opening. The distance between the LED cell and the detector should be 2 to 3 in (5.0 to 7.5 cm), while the IRED cell may require 6 in (18 cm).

1. Construct the circuit shown in Fig. 2-5, using a LED and photodetector (MFOD300). The LED and photodetector should be mounted through a piece of stiff paper on which 10° lines have been drawn. (See Fig. 2-6.) Reduce light to a minimum. The photodetector can be mounted on its own separate piece of cardboard, so that it will be free to move.
2. The detector should be placed on the 0° line and 2 in in front of the LED. Adjust the 10-kΩ potentiometer so that the digital multimeter (DMM) easily reads 2.5 V. This initial voltage reading will be maintained.
3. Move the detector to the $+40°$ line, and readjust the distance and potentiometer setting so as to maintain a 2.5-V reading. The detector should always focus on the center of the LED. Maintain the same DMM reading. Make a pencil mark on the paper showing the position of the front of the detector. The potentiometer setting should not be changed.
4. Move to each radiant line between $+40°$ and $-40°$, and note the light reading. A light radiation pattern will be plotted.
5. Measure the voltage V across the LED, and compute the current flow I_f.
6. When each line has been marked, remove the LED and replace it with the IRED. Maintain the same supply voltage, and repeat the measurement.

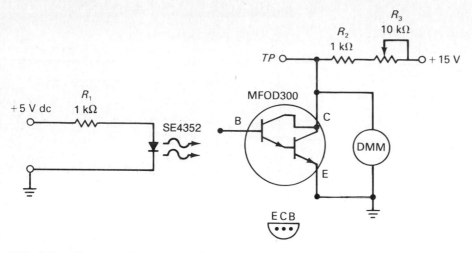

FIG. 2-5   Circuit arrangement, LED, and photodetector.

7. Start again at the 0° line, and adjust the detector distance so as to obtain the same detector output reference voltage as recorded for the LED (2.5 V).
8. Plot the locations on each line, and mark each spot with an X so that your points are not mixed.
9. When your plotting is completed, turn off the power. Connect the points which form the two radiation patterns.
10. Which emitter had the most concentrated and directed flux pattern?
11. What was the power input to the LED and to the IRED?
12. Measure the voltage V across the IRED, and compute the current flow I_p.

13. From the technical data provided, what is the emitted wavelength for the IRED?
14. How does your plotted radiation pattern compare with that provided for the tungsten lamp?

REVIEW QUESTIONS

On a separate sheet of paper, complete the following statements with an appropriate word or words.

1. Visible light, in the electromagnetic spectrum, is higher in frequency than _____ light.

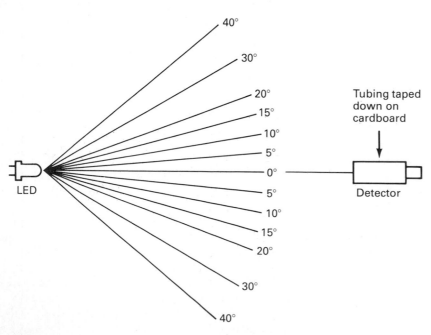

FIG. 2-6   Layout for angles in measuring LED radiation.

2. Ultraviolet light has a _____ wavelength than visible light.
3. An incandescent lamp has its greatest output in the _____ plane if the base is pointing down.
4. A LED is basically a(n) _____ junction diode.
5. The lumen is a unit of _____ measurement.
6. Adding a lens to the radiated output of a LED will _____ the beam.
7. To light, a LED must be _____-biased.
8. The output of a LED is measured in _____ .
9. The term *angstrom* relates to _____ .
10. The device which picks up the radiated light from a LED or IRED is called a _____ .

ANSWERS TO REVIEW QUESTIONS

1. Infrared
2. Higher
3. Horizontal
4. PN
5. Photometric
6. Narrow and intensify
7. Forward
8. Microwatts or milliwatts
9. Wavelength of light
10. Detector or photosensor

THE PHOTORESISTOR SENSOR

SCOPE OF STUDY

In this experiment the photoresistor and photovoltaic detector are considered. The photosensor output voltage is increased by use of amplifier(s).

OBJECTIVES

Upon completion of this experiment, construction of circuitry, testing, and evaluation of the data, you will be able to:

1. Describe the operation of photoresistors (photoconductive cells)
2. Use the photoresistors as controllers of power circuits
3. Utilize solar cells as circuit controllers of power circuits

BACKGROUND

In this discussion, two types of photoconductive cells are described, namely, the phototransistor (also referred to as the photoconductive cell) and the solar cell (also called the voltaic cell).

The photoresistor is generally made of either cadmium sulfide (CdS) or cadmium selenide (CdSe). The devices are made by the deposition of a layer of the semiconductor material on a substrate of ceramic or silicon. A clear coating of glass or plastic, to form a lens, can be used to focus the light. The semiconductor material, in a dark state, has few free electrons; however, when light (photons) irradiates the cell's surface, the electron flow increases and resistivity decreases. A dark cell may have a resistance of 30 to 50 MΩ while an illuminated cell's resistance may drop to under 5 kΩ. Dark-to-light resistance ratios of 10,000/1 are not unusual. Accompanying the resistance change is also a change in response time. The cell does not respond instantaneously to the influx of light.

The cells are sensitive to different wavelengths. The cadmium sulfide photocell peaks in the region of 0.60 μm (6000 Å). The cadmium selenide cell peaks in the region of 0.7 to 0.75 μm. Both peak above the visible light response.

Besides the composition of the semiconductor (which also includes selenium, germanium, and silicon), a number of geometric patterns are used to obtain the desired effects. Zigzag or interleaved patterns provide greater surface areas but a lower operating voltage. The photoresistor can be used as a potentiometer for the biasing of oscillators or amplifiers. Figure 3-1a shows the cell as a variable resistor connected in series with a load resistor R_L. Figure 3-1b shows the cell being used to change the bias on a transistor amplifier.

As the light increases on the cell in Fig. 3-1a, more current will flow through the load resistor R_L.

If the amplifier in Fig. 3-1b were biased to class A when the cell was dark, the amplifier would conduct and would go to cutoff when the cell was irradiated. If R_1 and the cell were interchanged, the transistor would be switched from off to on by the presence of light.

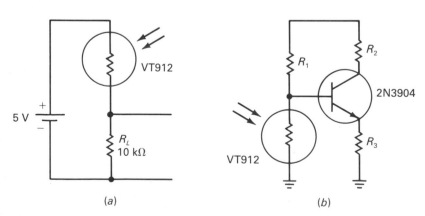

FIG. 3-1 **Using a photocell for biasing an amplifier.**

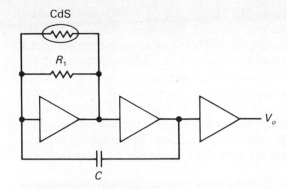

FIG. 3-2 Photoresistor-controlled oscillator.

The amplifier can also be used to frequency-modulate an oscillator. Then the frequency of oscillation is dependent on the level of light intensity present. Figure 3-2 shows a circuit for an oscillator which uses logic inverters. The frequency depends on the RC time constant. As the light increases on the cell, the value of R_1 decreases, hence the frequency of oscillation increases $[f = 1/(RC)]$.

Photoresistive cells are used in cameras for automatic exposure control, in street lighting, in counting or sensing changes for production control, and in measuring and determining the color density of blood. The components are easy to use, small, reliable, and economical. Figure 3-3 shows a table of typical photoresistive cells.

The VT912 has a resistance ratio of 20 MΩ/15 kΩ, or 1300/1. The cell is made of cadmium selenide (CdSe) with a peak response at 0.735 μm. The -6-dB points (down 50 percent) are 0.720 and 0.800 μm, and the rise time, to 62 percent of peak, is 2 milliseconds (ms).

The photovoltaic cell is a PN junction diode. The P material is often made of selenium or silicon, and the N material is cadmium or silicon.

Light irradiating the solar cell (junction area) reduces its energy band and causes electrons to move toward the N-type material while the holes move toward the P material. Externally, a dc potential can be measured which is in the range of 0.6 to 0.7 V, with the P material terminal being positive and the N material being negative.

The surface area determines the current supply capability. Cells can be connected in series to increase the total voltage or in parallel to increase the total current.

Part no.	Photosensitive material	Peak spectral response (Å)	2-fc Resistance (kΩ)	Minimum dark resistance 5 s after 2 fc	Maximum voltage
VT901	Type 0	5650	4.3	500 kΩ	100
VT902	Type 0	5650	8.0	750 kΩ	100
VT903H	Type 0	5650	44.0	5 MΩ	300
VT904L	Type 0	5650	88.0	10 MΩ	300
VT904	Type 0	5650	110.0	20 MΩ	300
VT912L	Type 1	7350	6.0	5 MΩ	100
VT912	Type 1	7350	15.0	20 MΩ	100
VT913H	Type 1	7350	32.0	50 MΩ	100
VT914L	Type 1	7350	64.0	100 MΩ	100
VT914	Type 1	7350	140.0	200 MΩ	250
VT93L	Type 3	5500	11.0	1 MΩ	100
VT931	Type 3	5500	22.0	1 MΩ	100
VT932	Type 3	5500	100.0	50 MΩ	100
VT941	Type 4	6750	2.0	500 kΩ	100

FIG. 3-3 Series VT900 light-dependent resistors. (*Courtesy of VACTEC Inc.*)

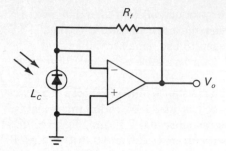

FIG. 3-4 Photocell amplifier.

Note that the early photovoltaic cells were made from a coating of selenium. The voltage output was smaller than that of the PN junction.

The photovoltaic cell has its peak response in the range of 0.5 μm. Cells made from indium antimonide operate in the near-infrared region. Most solar cells are very active in the visible light spectrum. The efficiency of the solar cell lies in the range of 5 to 15 percent. Figure 3-4 shows a solar cell with an amplifier. The voltage output of the amplifier is

$$V_o = i_c R_f$$

With a cell current of 5×10^{-6} A and $R_f = 100$ kΩ, the output voltage V_o is 0.5 V. Making the resistor larger or increasing the light radiation on the photocell will increase the output voltage. Since the signal voltage is applied to the negative input to the amplifier, the output signal will be inverted.

SUMMARY

The photoresistor, also referred to as the photoconductive cell or light-sensitive resistor, changes its resistance value in proportion to the irradiated light energy present. The resistance of the cell has a dynamic range of over 1000 to 1. The dark resistance can range from 10 to 100 MΩ.

Two types of materials are generally used in making photoresistors, namely, cadmium sulfide (CdS) and cadmium selenide (CdSe). The semiconductor material is deposited on a ceramic or silicon base using any number of geometric patterns.

The photovoltaic cell (solar cell) is a PN junction diode, which develops a potential of approximately 0.5 V across the junction when the cell receives light energy (photons). The cells can be connected in series for increasing voltage or in parallel for increasing current.

Both types of sensors can be used with amplifiers to record on-off light changes or light variations. Switches can be made to open or close, pinholes in sheet steel can be located, or light reflected from the end of a cassette tape can be made to stop the drive motor.

TESTS AND MEASUREMENTS

MATERIALS REQUIRED FOR EXPERIMENT

Active Devices	Others
VT912 Photoresistor	5-V Lamp
555 IC timer	8-Pin IC socket
TL081 (LF351) IC amplifier	**Capacitor** 0.001 μF
Resistors 10 kΩ (2) 100 kΩ 1 kΩ	

This experiment involves testing a photoresistor in several different circuits. The unit will be evaluated by itself, and it will be used to control the frequency of an oscillator. Since the device is sensitive to ambient light, it should be enclosed in black tubing, as in Experiment 2

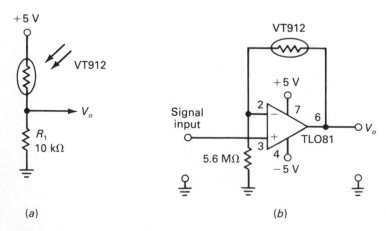

FIG. 3-5 Evaluating a photoresistor.

(see Fig. 2-6). A small tungsten lamp will be used as the light source. A solar cell will not be evaluated.

Photoresistor Evaluation

1. A 5-V lamp is used as a light source, or ambient light can be used. Construct the circuit shown in Fig. 3-5a. Measure the voltage across R_1 when the cell is dark, $V_{L(dark)}$, and when the cell is exposed to a light source, $V_{L(light)}$. With the voltage off, or the photoresistor out of the circuit, measure the resistance of the photoresistor when it is dark, R_D, and when it is illuminated, R_L.

Amplifier Evaluation/Gain Control

2. Construct the circuit shown in Fig. 3-5b. The cell is used in the feedback loop of the amplifier. As light impinges on the cell, its resistance decreases and so does its gain. The gain of the amplifier is determined by $A = R_f/R_1$.
3. Feed the amplifier with a 1-kHz sine wave signal at a 1-V level. When the cell is dark and when it is light, record the output voltage, $V_{o(dark)}$ and $V_{o(light)}$, respectively. Does light on the sensor cause a reduction in stage gain?

Oscillator Evaluation

4. Figure 3-6 shows the photoresistor used in an oscillator circuit. The frequency of oscillation depends on the amount of light radiation.
5. Record the frequency of oscillation when the cell is dark, $F_{o(dark)}$, and when the cell is light, $F_{o(light)}$.
6. Sketch the output waveshape V_o when the cell is illuminated.

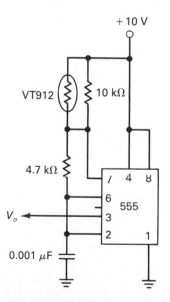

FIG. 3-6   Oscillator with photoresistor control.

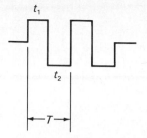

FIG. 3-7   Square wave.

7. Does the frequency increase or decrease with light?
8. A space capsule is circling the earth. How will you know when the craft is facing the sun?
9. Does the photoresistor control t_1 or t_2 in the sketch shown in Fig. 3-7?

REVIEW QUESTIONS

Test your comprehension of the subject by answering the following questions. Answer True or False on a separate sheet of paper.

1. A photoresistor exhibits a negative temperature coefficient when exposed to light.
2. The dark-to-light resistance ratio of a phototransistor is in the range of 10 to 1.
3. A light-sensitive resistor is generally made of cadmium sulfide or selenide.
4. The visible spectral response of a photoresistor is in the infrared range.
5. A solar cell is a photovoltaic cell.
6. A solar cell is formed by an NPN junction.
7. In a photovoltaic cell, the positive terminal is the P-type material.
8. When light falls on a solar cell, the electrons move toward the N-type material.
9. When a photoresistor is used in the feedback loop of an IC amplifier, the radiant will cause a decrease in amplifier gain.
10. The voltage output, no load, of a solar cell is in the range of 0.1 to 0.3 V.

ANSWERS TO REVIEW QUESTIONS

1. T
2. F
3. T
4. F
5. T
6. F
7. T
8. T
9. T
10. F

THE PHOTODIODE

SCOPE OF STUDY

In this experiment you will study the structure of a photodiode and its practical application as a light sensor.

OBJECTIVES

Upon completion of this experiment, construction of circuitry, testing, and evaluation of the data, you will be able to:

1. Describe the theoretical operation of the photodiode
2. Connect a photodiode sensor in a circuit to record the presence of, and changes in, light radiation
3. Record and describe how radiated light affects the current flow in a photodiode

BACKGROUND

Visible light or infrared emitters convert electricity to light. The action which causes light to be radiated in a LED is the movement of minority carriers as a result of a forward-biased current flow. The diode which radiates light, under proper conditions, could be made sensitive to light, and hence the junction could operate as a photosensor. When a silicon diode is reverse-biased, the electrons and holes move away from the PN junction (depletion area). Light of the proper wavelength radiating the junction causes hole-electron pairs to be formed, and this results in an increase in the flow of current to the outer circuit. The wavelength and sensi-

tivity of the device are related to the type of doping and the depth of penetration of the radiant light. Figure 4-1 shows a reverse-biased photodetector.

The power density of the radiated flux H (measured in milli- or microwatts per square centimeter) determines the current flow I_L. If no light radiation is present (zero light), a small leakage current called *dark current* I_D will remain. The amount of dark current depends on the reverse-biased voltage, the series or load resistance, and the ambient temperature.

The spectral response of the photodiode ranges from 400 to 1100 nm.

Silicon, PIN, photodiodes can be used as photovoltaic sources. Light impinging upon their depletion area creates a potential which is proportional to the incoming light. No external bias is required since the junction generates its own electromotive force. The photodiode is used in applications which include card reading, ambient light controls, slide projectors, and TV sets. The photodiode senses the presence of radiated energy and causes a control current to flow through external circuitry. Typically, the diode current flow is measured in the low microamperes.

Figure 4-2 shows the electrical characteristics of a typical photodiode. The spectral response is shown in Fig. 4-3.

RESPONSIVITY

The *responsivity* R is the ability of a photodiode to increase its back-biased current as a result of an increase in light. The responsivity of the photodiode is

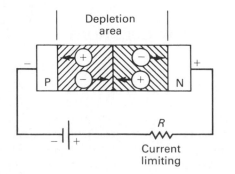

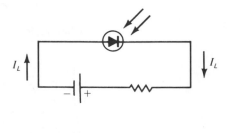

FIG. 4-1   **Reverse-biased photodiode.**

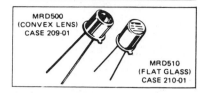

PHOTO DIODES
PIN SILICON
100 VOLTS
100 MILLIWATTS

MRD500
(CONVEX LENS)
CASE 209-01

MRD510
(FLAT GLASS)
CASE 210-01

... designed for application in laser detection, light demodulation, detection of visible and near infrared light-emitting diodes, shaft or position encoders, switching and logic circuits, or any design requiring radiation sensitivity, ultra high-speed, and stable characteristics.

- Ultra Fast Response – (<1.0 ns Typ)

- High Sensitivity – MRD500 (1.2 μA/mW/cm^2 Min)
  MRD510 (0.3 μA/mW/cm^2 Min)

- Available With Convex Lens (MRD500) or Flat Glass (MRD510) for Design Flexibility

- Popular TO-18 Type Package for Easy Handling and Mounting

- Sensitive Throughout Visible and Near Infrared Spectral Range for Wide Application

- Annular Passivated Structure for Stability and Reliability

TYPICAL OPERATING CIRCUIT

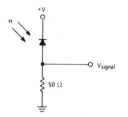

MAXIMUM RATINGS (T$_A$ = 25°C unless otherwise noted)

Rating	Symbol	Value	Unit
Reverse Voltage	V$_R$	100	Volts
Total Power Dissipation @ T$_A$ = 25°C Derate above 25°C	P$_D$	100 0.57	mW mW/°C
Operating and Storage Junction Temperature Range	T$_J$,T$_{stg}$	–65 to +200	°C

STATIC ELECTRICAL CHARACTERISTICS (T$_A$ = 25°C unless otherwise noted)

Characteristic	Fig. No.	Symbol	Min	Typ	Max	Unit
Dark Current (V$_R$ = 20 V, R$_L$ = 1.0 megohm; Note 2) T$_A$ = 25°C T$_A$ = 100°C	4 and 5	I$_D$	– –	– 14	2.0 –	nA
Reverse Breakdown Voltage (I$_R$ = 10 μA)	–	V$_{(BR)R}$	100	300	–	Volts
Forward Voltage (I$_F$ = 50 mA)	–	V$_F$	–	0.82	1.1	Volts
Series Resistance (I$_F$ = 50 mA)	–	R$_s$	–	1.2	10	ohms
Total Capacitance (V$_R$ = 20 V; f = 1.0 MHz)	6	C$_T$	–	2.5	4	pF

OPTICAL CHARACTERISTICS (T$_A$ = 25°C)

Characteristic		Fig. No.	Symbol	Min	Typ	Max	Unit
Radiation Sensitivity (V$_R$ = 20 V, Note 1)	MRD500 MRD510	2 and 3	S$_R$	1.2 0.3	3.0 0.42	– –	μA/mW/cm^2
Sensitivity at 0.8 μm (V$_R$ = 20 V, Note 3)	MRD500 MRD510	– –	S$_{(\lambda = 0.8 \mu m)}$	–	6.6 1.5	– –	μA/mW/cm^2
Response Time (V$_R$ = 20 V, R$_L$ = 50 ohms)		– –	t$_{(resp)}$		1.0		ns
Wavelength of Peak Spectral Response		7	λ_s	–	0.8	–	μm

FIG. 4-2 Photodiode characteristics of the MRD500.

rated in milliamperes per milliwatt at a certain wavelength. The Honeywell photodiode SE3452 has a rating of 0.5 mA/mW. When the radiated light at the cell is 2 mW, the diode will produce a current flow of 1 mA (0.5 × 2 mW). The peak response of the SE3452 is approximately 820 nm. Figure 4-4 shows a back-biased photodiode connected to a load resistor where a signal voltage is developed.

A current of 0.1 mA through a resistor of 10 kΩ will produce a signal voltage of 1 V. The higher the value of the load resistor, the higher will be the signal voltage, but the lower will be the frequency response. While the responsivity of the photodiode is in the range of 0.2 to 0.6, the phototransistor has a rating in the range of 100 and the photo-Darlington's rating is 400 to 600 because of the β multiplication factor.

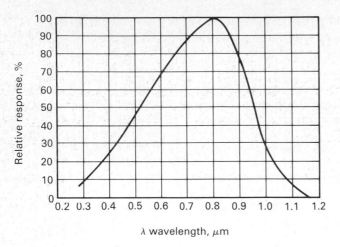

FIG. 4-3　Spectral response of photodiode MRD500.

The diode's capacitance is small, and hence its response time is small — 50 ns or less. The phototransistor's rise time is in the range of 1 to 20 μs. It is far slower than the photodiode. The Darlington is still slower, with a rise time of 10 to 100 μs. High-speed switching, therefore, requires the use of photodiodes. Its lower gain has to be made up by a high-gain, low-noise, wideband amplifier.

In evaluating light emitters and sensors, consideration has to be given to the rise and fall times of the device. The main characteristics of photodevices are the following:

1. Spectral range
2. Power output (emitter)
3. Sensitivity (sensor)
4. Response time
5. Mechanical considerations

SUMMARY

The photodiode is a reverse-biased silicon PN junction diode whose current flow depends on radiated light. The diode has a broad-wavelength response, although its current output is small. The diode's response is fast (small capacitance), and when used with transistors or IC amplifiers, the sensor makes an excellent controller for numerous applications.

TESTS AND MEASUREMENTS

MATERIALS REQUIRED FOR EXPERIMENT

Active Devices	Capacitors
Diode emitter SE4352	1.0 μF
Photodiode detector SD4478	47 μF
Resistors	
1 kΩ	
220 kΩ	

In the laboratory activity, the characteristics of a PIN silicon photodiode will be evaluated. An audio signal will be transmitted between an IRED exciter and a photosensor.

1. Construct the circuit shown in Fig. 4-5. Since the photodiode is sensitive to ambient light, it should be mounted in a black plastic tube or a piece of heat-shrink tubing about 1½ in long. The active side (junction) of the emitter should face the diode, and the separation between surfaces should be approximately 1 to 2 in. Both devices can also be mounted in sockets and then connected by a fiber-optic cable.
2. What are the values of the forward voltage V_f and current I_f of the near IRED? Measure the reverse voltage drop of the photodiode V_r and the current I_r.

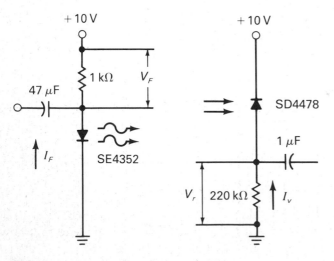

FIG. 4-4　Current flow from responsivity.

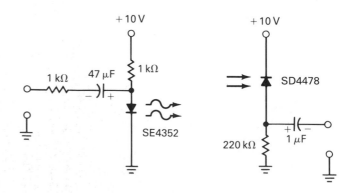

FIG. 4-5　Photodiode evaluation.

3. Close the light opening to the diode so that no light can enter. How much dark current flows? If you cannot measure it, refer to your technical data for a possible reason.
4. Feed a sine wave signal into the capacitor input which modulates the IRED current. Observe this waveshape on the oscilloscope. Use a frequency of 1 kHz. On a second channel, look at the output from the photodiode. Sketch both waveshapes, and record their amplitudes.
5. Does it make any difference if a square wave signal is used?

REVIEW QUESTIONS

On a separate piece of paper, complete the following statements with an appropriate word or words.
1. A photodiode is generally _____ -biased.
2. The wavelength of operations of a photodiode is dependent upon the _____ and the _____ of radiation into the junction.
3. The current flow of the photodiode, even with a strong radiant light, is measured in _____ .
4. Dark current is a measure of the _____ current.
5. The photodiode has a very _____ response.
6. The photodiode usually has a _____ wavelength.
7. The photodiode is generally made from a _____ chip.

ANSWERS TO REVIEW QUESTIONS

1. Reverse
2. Doping, depth
3. Microamperes
4. Leakage
5. Fast
6. Broad
7. Silicon

THE PHOTOTRANSISTOR

SCOPE OF STUDY

This experiment covers the basic characteristics and typical circuit arrangements for using phototransistors which have a greater sensitivity than the photodiode.

OBJECTIVES

Upon completion of this experiment, construction of circuitry, testing, and evaluation of data, you will be able to:

1. Describe the theory of operation of the phototransistor
2. Connect the photosensor into a circuit for measuring light changes
3. Compare the characteristics of this sensor with those of the photodiode

BACKGROUND

A *phototransistor* is a solid-state sensor whose collector and emitter current flow is directly related to irradiant rays impinging on its base. Any transistor will operate as a photosensor if the chip is exposed to light. The phototransistor, however, has some unique features which makes it more sensitive and respondent to certain wavelengths.

When light of the proper wavelength irradiates the base of the transistor, electron-hole pairs are formed, and these pairs constitute a flow of base current. The intensity of the current flow is directly related to the brightness of the light source. The main region of current generation is at the base-collector junction. The current flow in the base (because of light) appears as though an external diode existed between the base and collector. Figure 5-1 shows the basic circuit. The base lead connection may or may not be brought out for connection to an external circuit.

The emitter current, as in any transistor, is equal to $I_b(h_{FE} + 1)$; however, in the phototransistor the added diode current I_p must be considered. The equation then becomes $I_e = (I_p \pm I_b)(h_{FE} + 1)$, where h_{FE} is the forward dc current gain of the transistor. The base termi-

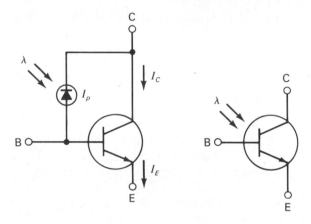

FIG. 5-1 **Phototransistor with reverse-biased diode.**

nal can be externally biased, and this will, in turn, affect the amplifier's gain.

The input impedance (resistance) of the transistor is approximated by $Z_{in} = R_{in} \times h_{FE}$.

Since the base is normally unterminated, R_{in} is higher than in a diode. The higher time constant results in a slower response time. In general, the higher the gain of the phototransistor, the slower is its response time. This is a most important parameter when the application is to switching circuits.

The frequency of the sensor can be rather broad. Figure 5-2 shows the spectral response of a Motorola MRD300. In order to increase the sensor's sensitivity, its base-to-collector junction is made large. The high h_{FE} and base current result in an increase in the dark leakage current I_{CEO}, where $I_{CEO(dark)} = h_{FE} \times I_{CBO}$. In turn, I_{CBO} is the collector-to-base leakage current. For the Honeywell SE3452, I_{CEO} is listed in the range of 4 to 8 μA. At room temperature, a typical current is 5 to 25 nA. The I_L value assumes a certain level of radiated flux density H.

The phototransistor can be connected into a circuit so that the output is taken from either the collector or the emitter. Figure 5-3a and b shows two arrangements of the circuit. Figure 5-3c shows how the phototransistor's current can be increased by adding another transistor. The emitter current of the transistor can be used as a current source for a second amplifier whose load is

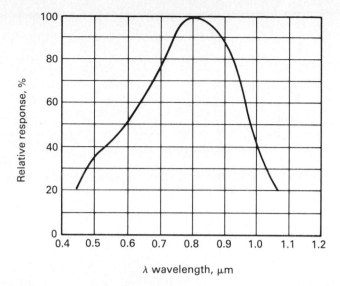

FIG. 5-2 Spectral response of phototransistor MRD300.

R_L. This load can be a resistor, relay, solenoid, or other current-activated device. The circuit is called a *photo-Darlington*.

SUMMARY

The NPN silicon phototransistor, with a wavelength ranging from near ultraviolet to infrared, has a collector and emitter current flow which increases with radiant energy on its collector-to-base junction. Compared with the photodiode, whose current flow is in microamperes, the phototransistor has a current flow in milliamperes. The increased current is due to the current gain

h_{FE}. The increased gain also increases the dark current and reduces the response time. In most applications, the base connection is not used.

TESTS AND MEASUREMENTS

MATERIALS REQUIRED FOR EXPERIMENT

Active Devices	Capacitors
SE4352 IRED	47 μF
SD3443-2 Phototransistor	1 μF (2)

Resistors
1 kΩ
10-kΩ potentiometer
47 kΩ
27 kΩ
10 kΩ
4.7 kΩ
5.6 MΩ

In the circuit evaluation that follows, a phototransistor will be irradiated with light and its characteristics observed.

1. Construct the circuit shown in Fig. 5-4. Since the phototransistor is sensitive to ambient light, it should be mounted into one end of a piece of plastic tube or heat-shrink tubing. The IRED is wired so that its light output can be varied. Sepa-

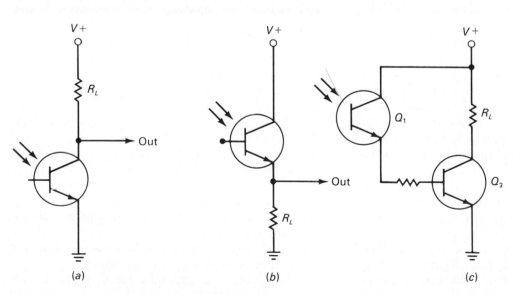

FIG. 5-3 Circuit wiring of the phototransistor.

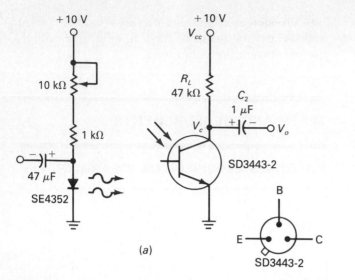
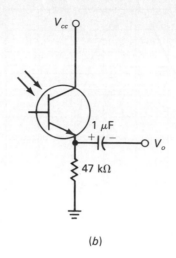

FIG. 5-4   Circuit arrangement for phototransistor evaluation.

rate the sensor from the IRED by approximately ½ to 1 in (depending on the ambient light).

2. Start with an R_L value of 47 kΩ. Disconnect the power to the IRED, and measure the dark voltage V_c. Calculate the dark current from $V_{cc} - V_c = V_{RL}$ and $I = V_{RL}/R_L$. What value of dark current do you calculate?

3. Reconnect and reduce the current flow in the IRED by resetting the 10-kΩ potentiometer until the IRED is essentially off. Record the voltage at the collector V_c. If it differs from the value found in Step 2, why?

4. Increase the current flow to the IRED. Observe the V_o of the transistor. As the IRED increases in current, what happens to V_c?

5. Adjust the 10-kΩ potentiometer until $V_c = ½V_o$, and reduce the value of R_L to 27, 10, and 4.7 kΩ. Record V_c for 47, 27, 10, and 4.7 kΩ.

6. With the 10-kΩ potentiometer adjusted as in Step 5 and $R_L = 4.7$ kΩ, feed a 1-kHz square wave into the 47-µF capacitor to the anode of the emitter. Connect the oscilloscope to V_o, with the external sync connected. Adjust the signal level until the square wave can be clearly observed at V_o. Sketch the pattern, and record the peak-to-peak voltage $V_{o(p\text{-}p)}$.

7. Try to reduce the gain of the phototransistor by connecting a 5.6-MΩ resistor between the base and ground, while observing the output. What happens to V_o?

8. Without changing the generator or IRED, turn off the power source and move R_L and C_2 to the emitter, as shown in Fig. 5-4b. Place a jumper between the collector and V+. Turn on the power and sketch $V_{o(p\text{-}p)}$.

9. From your sketches, determine whether the emitter output is in phase with the collector output. Are the output voltages the same?

10. Place the 47-kΩ resistor in the collector and measure V_c. Note that the output voltage is stable. Hold the phototransistor between your fingers for 1 full minute to see whether your body temperature will increase I_c just as radiant energy will. You can also use a heat gun or soldering iron to provide the heat source. After heating the transistor, note that the output voltage rises as the transistor cools.

REVIEW QUESTIONS

On a separate piece of paper, complete the following statements with an appropriate word or words.

1. The phototransistor has _____ gain than the photodiode.

2. The output from the sensor can only be taken from either the _____ or the _____.

3. The phototransistor has a(n) _____ wavelength response than a GaAs-emitting diode.

4. The collector current of the phototransistor is _____ than the photodiode.

5. Temperature causes the current flow in a phototransistor to _____.

6. Two transistors connected in tandem to increase the gain are called a _____ circuit.

7. Dark current is generally found to be in the _____ range.

8. The symbol H refers to _____.

9. The main light action in the device takes place at the _____ junction.
10. The phototransistor, when placed in the feedback loop of an operational amplifier, can control the _____ of the amplifier.

ANSWERS TO REVIEW QUESTIONS

1. More
2. Collector, emitter
3. Broader
4. Greater
5. Increase
6. Darlington
7. Microampere
8. Radiation flux density
9. Base-collector
10. Gain

EXPERIMENT **6**

DARLINGTON SENSORS

SCOPE OF STUDY

In this experiment you will study the characteristics and applications of high-sensitivity, narrow-bandwidth photosensors called the *Darlington sensors*. This type of detector has a higher gain, but a slower response time, than either the phototransistor or photodiode.

OBJECTIVES

Upon completion of this experiment, construction of circuitry, testing, and evaluation of data, you will be able to:

1. Describe the theory of operation of the Darlington-type phototransistor sensor
2. Connect the Darlington phototransistor in a circuit for evaluating its performance
3. Compare the characteristics of this sensor with those of the phototransistor

BACKGROUND

The Darlington amplifier consists of two cascaded transistors mounted on the same chip. The Darlington phototransistor provides more than twice the gain of the single unit. Figure 6-1 shows both the structure and the circuit analogy.

As noted, the light radiation and gain of the first stage are multiplied by the gain of the second stage.

The total gain is in the range of 10^3 to 10^5, and an approximation of gain is found from

$$I_{e2} = (I_{p1} \pm I_b)\, h_{fe1} \times h_{fe2}$$

With a higher gain, the speed of response becomes slower, and the dark current increases.

The Darlington may have a current-transfer ratio (CTR) of over 100, while the phototransistor has a CTR on the order of 10 to 20. A Darlington phototransistor is also incorporated in an optical coupler (see Experiment 7). The following is a list of some key characteristics of a Darlington sensor:

Dark current $I_{CEO} = 8.0$ nA
dc gain $h_{fe} = 15,000$
At a V_{CE} of 5 V, $I_c = 500\ \mu A$

Figure 6-2a shows a basic circuit in which a Darlington is used in a transmission system. Figure 6-2b shows the effect of various resistive loads on the frequency response of the sensor. Figure 6-2c shows how the load resistor is placed in the collector circuit. The larger the value of R_L, the lower the frequency response.

SUMMARY

The Darlington phototransistor is a two-transistor cascaded amplifier that is used as a photodetector. The gain of the detector is the product of the current gain of each transistor. Typical gains are in the range of 1500

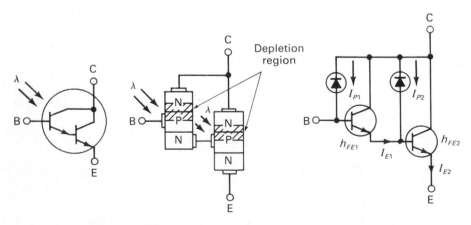

FIG. 6-1 Darlington phototransistor.

28

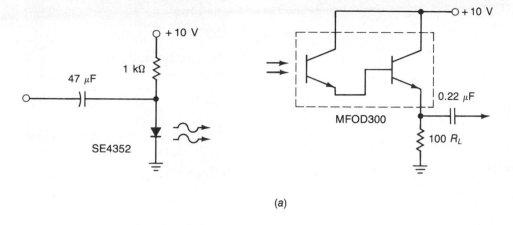

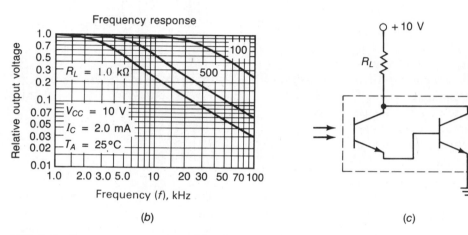

FIG. 6-2   **Frequency response of Darlington circuit.**

to 20,000. The higher gain results in a slower response and a higher dark current.

TESTS AND MEASUREMENTS

MATERIALS REQUIRED FOR EXPERIMENT

Active Devices	Capacitors
SE4352 IRED emitter	0.22 μF
MFOD300 Photo-Darlington sensor	47 μF
Resistors	
100 Ω	
470 Ω	
1 kΩ (2)	

In the circuit evaluation that follows, a Darlington phototransistor will be tested and evaluated.

1. Construct the circuit shown in Fig. 6-2a. Use a 100-Ω resistor for R_L. Connect the emitter and sensor by 50 cm of fiber-optic cable. The emitter and sensor have to be mounted in connectors for holding the cable. When the emitter is turned on, a voltage should appear across the load resistor.

2. *a.* Measure the voltage V_o across R_L when no light input is present. Remove the power from the IRED.

   b. Measure V_o when the IRED output is present without a modulating signal.

   c. What is the I_f of the IRED?

3. Modulate the IRED by using a 1-kHz square wave. Adjust the signal level so that a clean signal is seen at V_o. Make sure you do not overdrive the sensor. Also test by using a sine wave.

4. Observe the effects of R_L on frequency response by increasing the frequency of the function generator until the front end of the square wave equals $0.7V$ (where V is full voltage). Refer to Fig. 6-3. Sketch the waveshape present. At what frequency does this occur?

5. When rolloff of a square wave is observed, turn off the power and change R_L to 470 Ω. Again, sketch the waveshape for $R_L = 470$ Ω.

6. What happens to the waveshape as the value of R_L is increased?

29

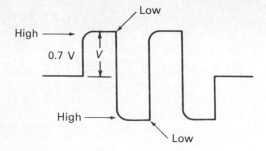

FIG. 6-3 High-frequency rolloff.

7. From previous experience, describe what you would expect to happen if the R_L were placed in the collector circuit and the output V_o were taken from the collector. Make the change and observe the output from the collector.

8. Compare the Darlington sensor with a phototransistor. Return to the 1-kHz signal, and place a 1-kΩ resistor in the collector of the Darlington circuit. What is the level of $V_{o(p-p)}$? Replace the Darlington sensor with a phototransistor; use the same emitter circuit as shown in Fig. 6-2a. What is the value of V_o? How much has the gain (Darlington/transistor) increased? Express it as a ratio.

REVIEW QUESTIONS

Test your comprehension of the subject by answering the following questions. Answer True or False on a separate piece of paper.

1. The gain of the Darlington phototransistor is higher since $h_{FE1} + h_{FE2}$ is present.
2. Because of the higher gain, the Darlington will have a higher frequency response than a PIN diode.
3. The dc gain of a Darlington is in the range of 1000 to 20,000.
4. The Darlington can be used at lower light levels than a phototransistor.
5. If the base of the photo-Darlington were grounded through a resistor, the gain would be decreased.

ANSWERS TO REVIEW QUESTIONS

1. F
2. F
3. T
4. T
5. T

SCOPE OF STUDY

In this experiment the characteristics and practical applications of optocouplers are studied. The device acts as a relay.

OBJECTIVES

Upon completion of this experiment, construction of circuitry, testing, and evaluation of the data, you will be able to:

1. Describe the characteristics and advantages of using an optocoupler in place of discrete components
2. Compute the current-transfer ratio (CTR)
3. Use and test couplers in practical circuits to determine how they perform with both pulses and analog signals

BACKGROUND

The *optocoupler,* also called an *optoisolator,* is a completely sealed IRED exciter and a photodetector. The exciter and detector are two completely isolated circuits; yet signals can be readily transferred between them. A low-voltage source can be made to control a high-voltage output circuit with complete isolation and without the high potential danger often encountered.

The coupler may contain an IRED emitter with a photodiode, phototransistor, Darlington, or laser sensor. To effectively use these devices, their characteristics must be known. Although many of their characteristics are similar to those for circuits using discrete components, there is a difference in the degree of isolation between the input and output circuits. The isolation features can be divided into three types: *isolation resistance, isolation capacitance,* and *dielectric breakdown ability.* The isolation resistance is in the order of 1×10^{11} Ω. The circuit resistance, external to the coupler, may be much lower in value. The capacitive isolation ranges from under 1 pF to under 3 pF and is the capacitance of the dielectric materials. Again, the circuit board layout may have a higher capacity than the coupler. Both the resistance and capacitive values are affected by the distance and medium between the

source and detector. A piece of glass, which is often used, affects the isolation characteristics.

The dielectric resistance breakdown, rated in volts, defines the maximum voltage that can be applied. Other factors such as waveshape, temperature, and altitude also affect the dielectric breakdown rating. Although a coupler may withstand 1000 V dc, it may only withstand 500 V ac. Couplers which have been previously subjected to high-surge voltages may exhibit a higher leakage resistance and/or short circuits between elements.

Input-output characteristics of a coupler, as previously mentioned, are similar to those for circuits which use discrete components. The input of the coupler is usually an IRED, and the output is one of several types of sensors. In some couplers, additional reflective surfaces may be added, or components such as an R or C may be needed in order to meet specific applications. The choice of IRED emitter to photodiode, phototransistor, or Darlington will depend on the specific application. A Darlington, for example, will produce a rather high output current of 10 to 100 mA; however, the input to the IRED may operate on only 0.3 to 3 mA. In such low current ranges, IREDs are not too effective. The high output currents can be attractive if the output saturation resistance is low, thus improving the drive to external elements. For a transistor output, the normal saturation voltage of V_{CE} is in the range of 0.3 to 0.5 V; but for a Darlington, $V_{CE(sat)}$ may range from 1.0 to 1.5 V. The saturation resistance is 2 to 4 times higher.

An important consideration in the use of the optocoupler or optoisolator is the current-transfer ratio. This parameter measures how much current is transferred from the IRED to the sensor in the presence of complete electrical isolation. The CTR, used with any LED-sensor combination, describes the current gain (or loss) from input to output. Essentially, the CTR is similar to comparing the I_C to I_B in a transistor circuit. Figure 7-1 shows the emitter-sensor circuits of a coupler.

The CTR in a coupler is defined as the ratio of I_C to I_F (see Fig. 7-1), and from the practical point of view, the combined circuit acts as a CE amplifier, except for the fact that there is no common tie between the input-output circuits. Depending on the type of sensor, the CTR can range from a loss to a gain of over 1000.

31

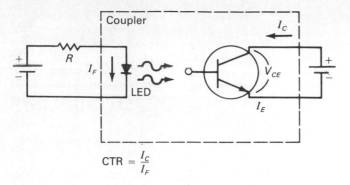

CTR $= \dfrac{I_C}{I_F}$

FIG. 7-1   Emitter-sensor circuits of a coupler.

Figure 7-2 shows the transfer curve of an IRED photodiode coupler. If a forward input current of I_F to the IRED is in the range of 10^{-3} and the output I_C from the sensor is in the order of 10^{-6} A, then the CTR will indicate a loss.

Figure 7-3 shows the transfer curves for a 4N25 coupler which uses a phototransistor output. With an input current of 10 mA (at 25°C), the ouput current I_C is 2 mA. The current transfer is operating at 20 percent.

Figure 7-4 shows two tables which list some Motorola couplers. Note that in a, for transistors, the CTR ranges from 2 to 225 percent. In b, for Darlingtons, the CTR ranges from 50 to 1000 percent.

Some couplers are designed for specific applications whereas others may be for general use. For example, a 4N37 is a LED emitter-to-phototransistor sensor with a 100 percent minimum CTR (when $I_F = 10$ mA and $V_{CE} = 10$ V). Also $V_{CE(sat)} = 0.3$ V ac at $I_F = 10$ mA and $I_C = 0.5$ mA. The MOC5010 is an optically coupled ac linear amplifier, where a diode sensor and oper-

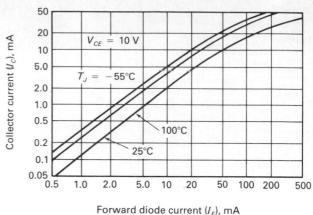

FIG. 7-3   Transfer characteristics of 4N25/4N26.

ational amplifier (op amp) are included within the coupler.

SUMMARY

The optical coupler (isolator) enables the circuit designer to control output circuits by varying the energy emission of a LED or IRED. The input-output circuits are electrically separated by a high resistance. The coupler can contain a photodiode, transistor, Darlington transistor, op amp, gate, or other device needed in interfacing with the controlled circuit. The current-transfer ratio between the input and output is referred to as CTR and is similar to the h_{FE} characteristic of a transistor.

TESTS AND MEASUREMENTS

MATERIALS REQUIRED FOR EXPERIMENT

Active Devices
4N37/TIL111/TIL114 Coupler,
  phototransistor output
4N30/TIL119 Coupler, Darlington output
MOC5010 Coupler, linear output

Resistors	Capacitors
470 Ω	47 μF
820 Ω	1 μF
1-kΩ Potentiometer	

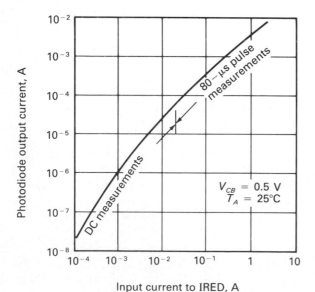

FIG. 7-2   Transfer curve (CTR).

In this experiment, three types of couplers are tested and evaluated. In later experiments, other types of couplers will be used in experimental control circuits.

Device type	Current-transfer ratio, dc, % min	$V_{(BR)CEO}$, V min
TIL112	2.0	20
TIL115	2.0	20
IL15	6.0	30
MCT26	6.0	30
TIL111	8.0	30
TIL114	8.0	30
MOC1006	10	30
IL12	10	20
4N27	10	30
4N28	10	30
H11A4	10	30
TIL124	10	30
TIL153	10	30
IL74	12.5	20
MOC1005	20	30
TIL125	20	30
TIL154	20	30
4N25	20	30
4N26	20	30
H11A2	20	30
H11A3	20	30
H11A520	20	30
IL1	20	30
MCT2	20	30
TIL116	20	30
4N38	20	80
H11A5	30	30
MCT271	45	30
H11A1	50	30
H11A550	50	30
TIL117	50	30
TIL126	50	30
TIL155	50	30
CNY17	62	70
MCT275	70	80
MCT272	75	30
MCT277	100	30
4N35	100	30
4N36	100	30
4N37	100	30
H11A5100	100	30
MCT273	125	30
MCT274	225	30

(a)

Device type	Current-transfer ratio, dc, % min	$V_{(BR)CEO}$, V min
4N31	50	30
H11B3	100	25
4N29	100	30
4N30	100	30
MCA230	100	30
H11B255	100	55
MCA255	100	55
H11B2	200	25
MCA231	200	30
MOC119*	300	30
TIL119*	300	30
TIL113	300	30
MOC8030*	300	80
TIL127	300	30
TIL128*	300	30
TIL156	300	30
TIL157*	300	30
H11B1	500	25
4N32	500	30
4N33	500	30
MOC8020*	500	50
MOC8050*	500	80
MOC8021*	1000	50

* Pin 3 and pin 6 are not connected.

(b)

FIG. 7-4   Transfer gains of Motorola couplers. Isolation voltage is 7500 V (min) on all devices. (*a*) Transistor output. (*b*) Darlington output.

Optocoupler/Phototransistor

1. The 4N37, TIL111, or similar devices will be evaluated initially by using the circuit shown in Fig. 7-5.

NOTE: Two separate power supplies would be used in an actual application. For experimental purposes, if your power supply has a common ground, terminals 2 and 4 are grounded. If floating power supplies are available, they should be used. The high-voltage isolation capability of the coupler is not being evaluated; therefore, a common ground can be used.

2. Determine the CTR of the device by measuring V_c, I_c, and the value of I_f through the IRED.

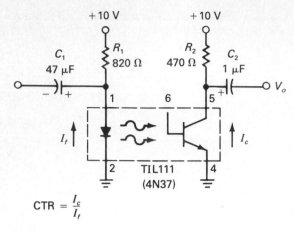

$$CTR = \frac{I_c}{I_f}$$

FIG. 7-5 Evaluation of a 4N37/TIL111.

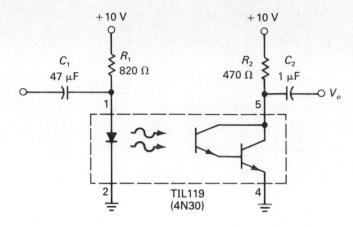

FIG. 7-6 Optocoupler using a Darlington output.

3. Set your function generator so as to produce a positive square wave pulse output at approximately 1 kHz. Adjust the amplitude while observing V_o. Do not overdrive the sensor. Sketch the output, and record the output $V_{o(p\text{-}p)}$ as well as the input.

4. Change your pulse polarity to negative-going, and again sketch the p-p output and input voltages.

5. Increase the frequency, and determine the 3-dB point $f_{3\,dB}$ (where the leading edge rolls off to 0.7 times the maximum height of the signal).

Testing the 4N30/TIL119/Darlington Output

6. Connect the circuit shown in Fig. 7-6.

7. Find I_f and I_c, and compute the CTR.

8. Determine whether the 4N30 has a lower frequency response by repeating the pulse test (Step 5).

9. Use a 1-kHz sine wave at the input, and adjust your level so that V_o is not saturated. At normal peak-to-peak (p-p) output, measure both the gen-

erator's output (p-p) and V_o. What is the value of the gain or loss?

Testing the MOC5010 Linear Output Coupler

10. Construct the circuit shown in Fig. 7-7.

11. Use a signal input of a sine wave, 1 kHz, and adjust its level so that V_o is at maximum yet undistorted. What is the ac gain between V_o and the generator output (input to emitter)?

12. Determine the frequency response of the optocoupler, in hertz, between the lower and upper 3-dB points.

13. Rather than lowering the signal voltage source, try to adjust the gain by reducing the modulation level. Add a 1-kΩ potentiometer in parallel with the IRED (see Fig. 7-7). Other tests you might choose to make are:

  a. Determine the output impedance of the sensor amplifier

  b. Measure the overall distortion of the system

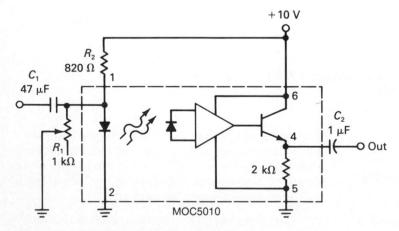

FIG. 7-7 AC linear output coupler circuit.

REVIEW QUESTIONS

Test your comprehension of the subject by answering the following questions. Answer True or False on a separate piece of paper.

1. If you do not forward-bias a LED, you cannot modulate it with a low-level signal.
2. A phototransistor output in an optocoupler has a lower saturation voltage than a Darlington output.
3. The 4N30 has a higher gain than a 4N37 coupler.
4. The MOC5010 coupler contains a photodiode output.
5. The CTR concerns the transfer loss in a coupler.
6. The voltage isolation between the input and output is between 1 and 3 kV.
7. The resistance isolation between the LED and the output is better than 10 MΩ.
8. The capacitive isolation in a coupler is less than 3 pF.
9. The optocoupler is used as a power device controller only.
10. The CTR of an optocoupler is always less than 1.

ANSWERS TO REVIEW QUESTIONS

1. T
2. T
3. T
4. T
5. F
6. F
7. T
8. T
9. F
10. F

LASCR SENSORS AND TRIAC COUPLERS

SCOPE OF STUDY

This experiment covers the application of special couplers used in power control. These couplers include both light-activated silicon controlled rectifiers (SCRs) and triacs.

OBJECTIVES

Upon completion of this experiment, construction of circuitry, testing, and the evaluation of the data, you will be able to:

1. Describe the operation of a light-activated silicon controlled rectifier (LASCR)
2. Use the LASCR in a circuit to control a power device by means of light activation
3. Use a driver-coupler to control a higher-power triac

BACKGROUND

The LASCR uses radiant light to activate its gate. The assembly consists of two transistors—a high-voltage PNP and a gate-controlled NPN transistor. Figure 8-1 shows how the circuit is arranged.

When the PN junction (diode) is light-activated, a current I_p flows into the NPN transistor and biases it on; this, in turn, activates the base of the PNP transistor, turning it on. The NPN transistor is required to be a high-gain amplifier circuit, since the light-sensitive photodiode produces a very small amount of current. Because of this high sensitivity, the LASCR is also sensitive to temperature and the applied line voltage. Most LASCR circuits are low-current-rated (under 2 A) and so are used as controllers for higher-power SCR equipment.

The LASCR, once activated, works the same as a discrete component (SCR). The current can be controlled, but the circuit must be deactivated (the anode or cathode line interrupted) to turn it off. You should refer to an SCR handbook or electronics text for a more detailed description of their operation in half- and full-wave circuits.

The triac, like the SCR, is a power controller. The triac, however, is controllable from almost zero to full power. It is ideally suited to variable-power drives, such as electric drills, heaters, light dimmers, and numerous appliances.

The high-power triac, in discrete-component form, requires a smaller-powered triac to control it. This low-power two-terminal unit is referred to as a *diac*. In the optical coupler field, this device is called a *triac driver*, or *bilateral switch*. The Motorola MOC3011 is one such driver (coupler). Figure 8-2 is a basic sketch of the device.

The LED (made of GaAs) operates on a forward voltage of 1.5 V at 10 mA. Its maximum current is 50 mA. Nominally 15 mA of diode current is adequate.

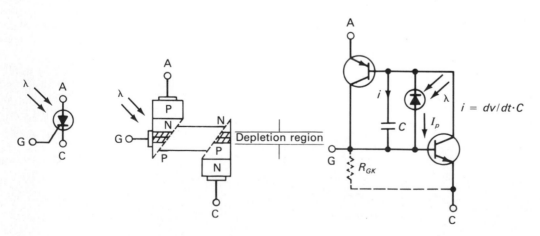

FIG. 8-1 LASCR construction.

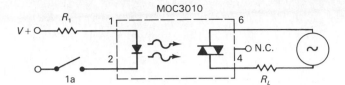

FIG. 8-2   Triac driver (3010/3011).

The detector has a blocking voltage (in the off state) of 250 V. In the on state, the detector conducts in either direction (both halves of the ac cycle) and will pass 100 mA with a voltage drop of 3 V.

Once it is triggered on, the detector's current can be controlled, and the detector will conduct as long as the current remains above 100 μA. At lower current levels, the detector returns to the blocked position. A current through the LED in the range of 10 mA will trigger the detector to the on state.

Figure 8-3 shows how the driver is used to control an external triac and a resistive load. The 180-Ω resistor prevents surges from entering the bilateral switch of the coupler. A logic gate is used in the circuit to turn on the LED. If the LED has a supply of +5 V, then $R_1 = 230 \ \Omega$ would be adequate. A standard value of 270 Ω would provide a flow of approximately 13 mA.

When the load is inductive and a phase shift exists between voltage and current, the triac may want to turn on with any sudden rise in voltage. This can be prevented by the addition of an *RC* circuit, which is referred to as a *snubber network*. Figure 8-4 shows the "snubber" formed by R_3-C_1. A hex buffer is used to turn on the coupler circuit.

NOTE: The value of R_1, which controls the LED currents, will vary depending on the type of hex buffer used.

SUMMARY

The LASCR is an SCR device which can be activated by the presence of light. The detector is available as a separate element or as part of an optocoupler.

The triac driver is used to control a high-power triac device. The coupler can be operated in the on-off mode or in a variable mode (by varying the light level).

TESTS AND MEASUREMENTS

MATERIALS REQUIRED FOR EXPERIMENT

Active Devices
MOC3010 Triac driver
2N6342A Triac or equivalent (6 A, 200 V)

Resistors	**Other**
47 Ω	PR15 Lamp, 5 to 6 V
220 Ω	
1-kΩ Potentiometer	

In this experiment you will use the triac driver to control a higher-power triac and its control of the intensity of a lamp. An ac voltage of less than 115 V is used because of safety reasons.

PROCEDURE

1. Construct the circuit shown in Fig. 8-5. A few basic tests, using a triac, will be performed.
2. The LED is arranged so that its intensity can be varied. Turn on the power. Close switches 2b, 3b,

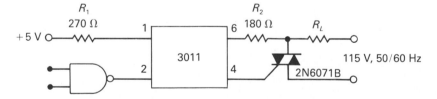

FIG. 8-3   Triac controller with resistive load.

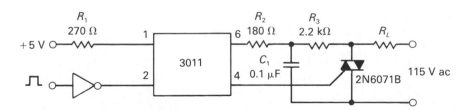

FIG. 8-4   Triac circuit with snubber circuit.

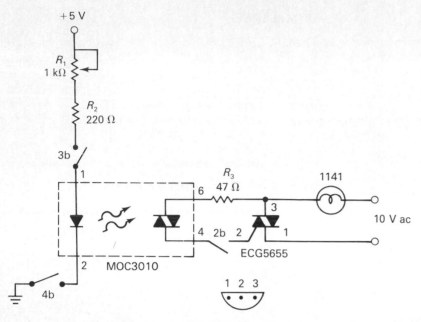

FIG. 8-5 Experimental triac control circuit.

and 4b, and vary the 500-Ω potentiometer. Does the lamp change its intensity as the potentiometer is varied?

3. From the voltage drop across R_1 and R_2 determine the minimum level of LED current I_f at which the ac lamp will no longer turn on. Is this current near the suggested 10-mA range?

4. Connect an oscilloscope across terminals 1 and 3 of the triac, and observe its waveshape as it turns on and is varied.

NOTE: Your oscilloscope is normally earth-grounded. It may be necessary to float the line cord in order to see the waveshape.

Sketch the waveshape observed at off, ½ on, and full on.

REVIEW QUESTIONS

On a separate piece of paper, complete the following statements with the appropriate word or words.

1. The LASCR uses an NPN transistor for the gate and a(n) _____ transistor for the control of the ac circuit.

2. Once turned on, the SCR circuit must be _____ to be turned off.

3. The triac controller is _____ over a wide range.

4. The triac drive is used to control a larger _____.

5. A snubber is used to correct ac problems relating to _____.

6. The triac driver can be operated in either the _____ or the _____ mode.

7. In a LASCR coupler, the detector used is a(n) _____.

8. The triac driver is similar to the _____, but the coupler also provides electrical isolation.

9. The triac coupler has a standard off voltage rating of _____ V.

10. The maximum LED current is _____ mA.

ANSWERS TO REVIEW QUESTIONS

1. PNP
2. Interrupted
3. Variable
4. Triac
5. Phase shift
6. On-off, variable
7. Photodiode
8. Diac
9. 250
10. 50

PART THREE

APPLICATIONS TO POWER

SCOPE OF STUDY

A number of experiments are provided in this section to show how the LED and sensors can be used to control power circuits.

OBJECTIVES

Upon completion of the study and experiments in this part and the evaluation of data, you will be able to:

1. Describe the operation of each power control circuit tested
2. Suggest applications for the principles involved in the circuits
3. Test and evaluate circuits which utilize couplers in high-voltage control applications
4. Communicate your understanding of optoelectronic components to other technical and nontechnical people
5. Research optoelectronic components for other applications
6. Maintain and repair power control equipment which uses optodevices
7. Read and comprehend technical articles which describe the use of optodevices in power-controlled circuits

BACKGROUND

In this part, a collection of circuits is introduced, including optodevices and other electronic components. The circuits include on-off, variable level, density control, motor speed control, and basic digital controllers. Only typical basic circuits are presented. Numerous variations in circuit design are possible for specific applications.

For each circuit, a components list and suggested testing procedures are provided. While 120 V, 240 V, or higher ac voltages could be controlled, the experimental circuit will require only 5 to 12 V ac. At the lower voltage level it is safer to work, and yet basic control principles can still be evaluated.

Some circuits incorporate optocouplers whereas others make use of LEDs and photosensors. Where couplers are used, lower-voltage-isolation devices have been specified. In actual applications, higher-voltage-isolation ratings are required.

THE ISOLATED ON-OFF CONTROLLER

SCOPE OF STUDY

This experiment covers the application of an optical coupler as a controller for turning power switches on and off.

OBJECTIVES

Upon completion of this experiment, construction of circuitry, testing, and evaluation of data, you will be able to:

1. Describe the operation of the on-off controller
2. Use an optocoupler to control a LED
3. Suggest applications for the use of an on-off controller

BACKGROUND

The isolated on-off controller makes use of an optocoupler. When a light source is turned on or off, de-pending on the sensor circuit, the relay controller can be made to close or open a higher-voltage controlled circuit. Figure 9-1 shows the experimental circuit to be evaluated. When light is present in the coupler, the phototransistor is illuminated and its resistance goes low.

The low value of resistance increases the base current of the control transistor (2N3904). The increased base current, times h_{FE}, causes the collector current to increase, and so the relay closes. A LED is used as a relay-closing indicator. Obviously, a higher-voltage relay could be used to control a motor or other power device. The coupler provides complete isolation between the controller LED and the power-controlled circuit.

Figure 9-1 shows a circuit for an on-off controller which activates the relay when the light goes on. In Fig. 9-2 the relay closes when the light goes off rather than when it turns on.

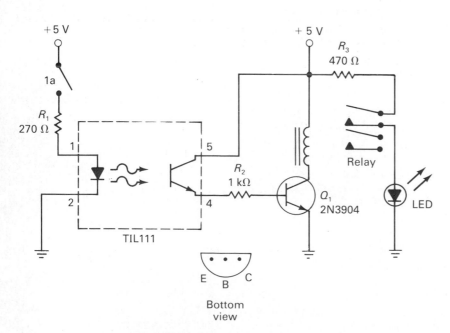

FIG. 9-1 On-off controller in which relay closes with light.

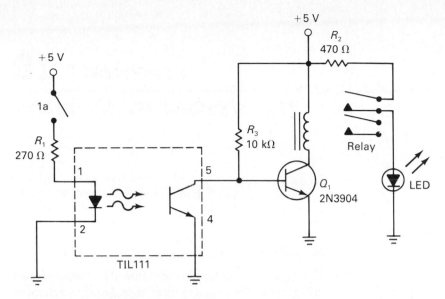

FIG. 9-2 Relay closes with no light.

TESTS AND MEASUREMENTS

MATERIALS REQUIRED FOR EXPERIMENT

Active Devices
4N37/TIL111 Coupler, phototransistor
2N3904 Transistor
LED, visible light

Resistors	Miscellaneous
470 Ω	Switch SPST
270 Ω	Relay 3 to 5 V dc
1 kΩ	
10 kΩ	

Two types of circuits are tested. One relay circuit is activated when light is present, while the second circuit is activated when no light is present. Both circuits use an optocoupler for isolation between the controlling and the controlled circuits.

1. Construct the circuit shown in Fig. 9-1.
2. The relay is operated as an on-off controller. Does closing switch 1a cause the relay to close?
3. Record the voltage at the base of Q_1, when switch 1a is open (no light) and when switch 1a is closed. What is the base voltage V_{BE} when 1a is open? When 1a is closed, what is V_{BE}?
4. What is the voltage drop across the relay when switch 1a is closed? Open?
5. Change the circuit arrangement to that shown in Fig. 9-2.

6. When no light is present, does the relay close?
7. In this circuit the phototransistor _____ the Q_1 base current to ground.
8. The optocoupler serves as a(n) _____.
9. Is the relay circuit isolated from the light source circuit?

NOTE: The circuit shows the use of a common ground; however, with two floating power supplies, the input and output circuits would be isolated.

10. What component determines the level of ac voltage and current that the circuit can control?

REVIEW QUESTIONS

On a separate piece of paper, complete the following statements with the appropriate word or words.
1. The nominal current flow through the LED in the coupler is _____ mA.
2. The resistor R_1 controls _____.
3. The sensor in the coupler is a photo_____.
4. The sensitivity of the relay is determined by _____.
5. Does ambient light have any effect on the operation of the coupler? _____

ANSWERS TO REVIEW QUESTIONS

1. 10
2. LED current
3. Transistor
4. The gain of the driving transistor
5. No

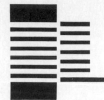

THE FLAME-OUT MONITOR

SCOPE OF STUDY

In this experiment an electric-powered furnace is controlled by a photosensor which monitors the flame's amplitude and/or color.

OBJECTIVES

Upon completion of this experiment and the evaluation of data, you will be able to:

1. Describe the operation of a flame-controlled furnace
2. Suggest applications for the principles involved in the control circuit
3. Test and evaluate a basic circuit similar to those used in industry
4. Communicate your understanding of the power control circuit to other technical and nontechnical people
5. Research applicable couplers for similar applications
6. Maintain and repair industrial/commercial equipment using optodevices, providing additional information and training is provided on the overall operation of the specific equipment

7. Read and comprehend technical articles which describe the use of similar optical devices

BACKGROUND

The flame-out monitor is a variation of the circuit shown in Fig. 10-1. The alarm or monitor indicator is activated when the light (from the flame in an oven or furnace) is extinguished. For example, the circuit might be used to activate a solenoid valve which turns off the fuel oil supply to the furnace if the flame does not start. Figure 10-1 shows a basic circuit in which relay contacts are used to control the gate current to an SCR. Switch 2a resets the SCR once it is in a conducting mode. The lamp, activated by switch 1a, simulates the furnace flame.

The relay contact's closure is used to provide the necessary gate current to the SCR. If an SCR requiring a lower gate current is used, the phototransistor could control the SCR directly, thus eliminating the need for the relay, its drive transistor, and associated components. Figure 10-2 shows a phototransistor circuit in which the sensor directly controls the gate current of an SCR.

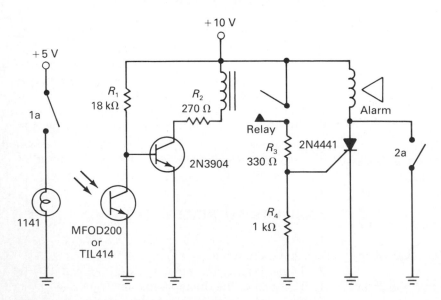

FIG. 10-1   Relay-controlled, SCR flame-out monitor.

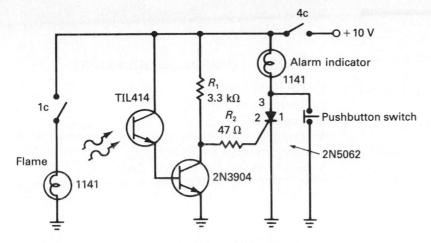

FIG. 10-2 Phototransistor control of an SCR.

TESTS AND MEASUREMENTS

MATERIALS REQUIRED FOR EXPERIMENT

Active Devices
TIL414 Phototransistor
2N3904 NPN transistor
SCR thyristor 2N5062
LED indicator

Resistors	**Other**
47 Ω	Switch SPST (2)
3.3 kΩ	Switch pushbutton SPDT
	Lamp type 1141, 12 V (2)

Two types of control circuits are evaluated. One circuit uses a relay to control an SCR while in a second circuit a photosensor directly drives an SCR.

1. Evaluate the circuit shown in Fig. 10-2. The alarm monitor could consist of a visual and/or audible alarm. Connect the +10 V dc (close 1c, 4c), and ground connections to the indicated terminals.
2. The control lamp should glow when the light source (furnace flame) is extinguished. Does the controlled indicator go on when switch 1c is opened?
3. When the flame light is on, record the voltage at the gate of the SCR, V_G.
4. Measure the voltage drop across R_2 when the alarm light is on. Open switch 1c to determine the gate current. Compute the gate current from $I_G = V_{RZ}/R_2$.
5. How much gate current I_G flows when the alarm indicator is off? Once the SCR conducts, reducing the gate current will not turn off the SCR. The SCR's anode current must be momentarily removed either by opening the anode or cathode circuit or by short-circuiting the anode to cathode via pushbutton (as shown in Fig. 10-2).
6. Turn the flame light on and off. Does the SCR also turn on and off?
7. Momentarily close the pushbutton switch while the flame light is on. Does the alarm light (alarm) go off?
8. From the technical data section in Appendix B, determine the typical current for phototransistor TIL414 when it is light-irradiated.
9. The experimental circuit can be used to monitor the presence of a flame in a furnace. The light from molten metal or an open flame can be the light source. The SCR gate current and the SCR control of the alarm indicator are a function of the change in transistor current. Record the collector voltage of the transistor for a no-flame condition, and when the flame is present.
10. The experimental circuit operates in a manner which is similar to a solid-state _____ .
11. Besides turning on an alarm and indicator, the SCR could also start a motor or turn off a fuel valve. The voltage and current of which component determine the circuit's applications?
12. To turn off an SCR once it is conducting, the anode-cathode current must be _____ .
13. In a chemical processing plant, the liquid level in a tank is to be maintained. Suggest a circuit diagram.

REVIEW QUESTIONS

On a separate piece of paper, complete the following statements with the appropriate word or words.

1. Could a flame detector circuit be made to respond to the color of the flame? How?
2. Once an SCR is fired on, it stays on until _____.
3. A TIL414 is a photo_____.
4. The sensitivity of the controller is determined by _____.
5. The voltage and current rating of an SCR is determined by the _____.

ANSWERS TO REVIEW QUESTIONS

1. Yes, with filter(s).
2. Its current is interrupted.
3. Transistor
4. The gain of the phototransistor, or the transistor-gate current of the SCR
5. Size of the load to be controlled

THE OPTICAL LOGIC CONTROLLER

SCOPE OF STUDY

This experiment demonstrates how two photosensors can be used to provide the logical AND function. Two events must take place for an alarm to sound or a control to be activated.

OBJECTIVES

Upon completion of this experiment, construction of circuitry, experimental testing, and evaluation of the data, you will be able to:

1. Describe the operation of a logic-controlled power controller
2. Suggest applications of the principles involved in the experimental circuit
3. Test and evaluate circuits similar to the one presented in this experiment
4. Communicate your understanding of the function of a logic-controlled power device to technical and nontechnical people
5. Research optoelectronic sensors for use as a power controller
6. Maintain and repair industrial/commercial equipment using logic-operated sensors, given additional information and training on the overall operation of the equipment

7. Read and comprehend technical articles which describe the use of similar devices

BACKGROUND

Industrial applications may require that two operations be controlled at the same time. For example, a conveyor belt must be moving AND products must be on the belt. The alarm will not sound if subsystem A AND subsystem B are both functioning. If either or both A and B are not operational, the alarm is sounded and production stops.

The operation may require the use of one or two different light sources to control the two photosensors. A wide variety of logic circuits can be formed, and this experiment demonstrates only one typical application. In this experiment two control circuits are investigated, namely, when A and B are on, the output is on and when A and B are off, the output is off. The truth table shown in Fig. 11-1 shows that only when A and B are off is the output on.

Figure 11-1 shows a circuit arrangement where both A and B sensors must be illuminated in order for the alarm to be on. The collector of Q_3 goes high for only the one condition (when A and B are on). This increase in collector voltage provides the gate current to the SCR. For all other conditions of A and B the circuit does

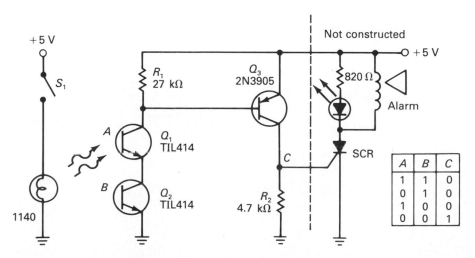

A	B	C
1	1	0
0	1	0
1	0	0
0	0	1

FIG. 11-1 When A and B are zero, the alarm sounds.

not function. In this experiment another type of logic control circuit is illustrated.

TESTS AND MEASUREMENTS

MATERIALS REQUIRED FOR EXPERIMENT

Active Devices
TIL414 (2) Phototransistor
2N3904 NPN transistor
2N3905 PNP transistor

Others	**Resistors**
6-V Lamp, type 1140	4.7 kΩ
Switch SPST	10 kΩ
	27 kΩ

Two types of logic circuits are evaluated, and the results are compared with a truth table.

1. Construct only the circuit shown to the left of the dotted line in Fig. 11-1. Place the lamp so that its light will illuminate both phototransistors. Ambient light can also be used. The circuit does not include the SCR or alarm lamp. Once the voltage across the 4.7-kΩ resistor goes high, it could be used to control the SCR and alarm. The output of Q_3 will be observed. Test the circuit and prepare a logic truth table. The output voltage at the collec-

tor of Q_3 should be high only when sensors A and B are illuminated with a bright light. When A and B sensors are on, the transistor Q_3 saturates and the voltage across the 4.7-kΩ resistor goes positive (high). This positive voltage supplies the gate current for the SCR.

2. Block the light from only sensor B, and record the voltage V_c at the collector of Q_3.
3. Block A and not B. What is the V_c of Q_3? Block A and B. Now what is the V_c?
4. Allow the light to radiate on both A and B, and record the collector voltage of Q_3.
5. From the voltage measurement V_c and the 4.7-kΩ resistor, determine the current flow I_c.
6. Is there a sufficient voltage to turn on the gate of an SCR? Figure 11-2 shows the logic control elements for a light controller in which the output is zero only when both inputs A and B are illuminated. When A is illuminated, its resistance goes low and the voltage across R_1 increases. This increase in voltage does not cause an increase in current to Q_3, since sensor B exhibits a high resistance when it is off and the current flow into the base of Q_3 is insufficient to saturate Q_3. Under this condition the collector of Q_3 remains high. This verifies the second statement in the truth table.
7. Answer statement 3 in the truth table. Does the voltage drop across R_1 increase when B is illuminated?
8. Answer statement 4. Does Q_3 have a large or small base current when neither A nor B is illuminated?
9. When both A and B are illuminated (statement 4), the base current of Q_3 has _____ and the collector voltage has _____ .

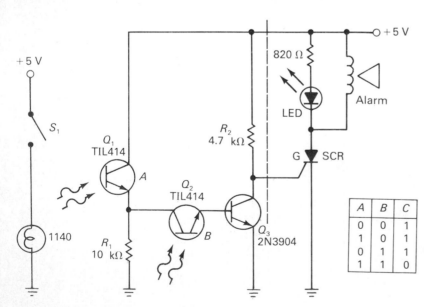

FIG. 11-2 Logic light controller.

10. Is the following statement true or false? A typical application of AND logic is expressed as follows: A metal shear is activated only when the metal is in place and the operator's fingers are in a safe position.

11. When the level of the liquid in the dye vat and the concentrate of the mixture are correct, the tank can be emptied. Which circuit would be best used, that in Fig. 11-1 or 11-2?

REVIEW QUESTIONS

On a separate piece of paper, complete the following statements with the appropriate word or words.

1. An AND circuit exists when two sensors _____.
2. In an OR circuit, either sensor can be _____.
3. In a punch press, hand injury could be prevented if a(n) _____ circuit was used.
4. A truth table describes _____.
5. The main power controller in Fig. 11-2 is the _____.
6. In the two circuits shown in Figs. 11-1 and 11-2, is the on-off operation gradual or do they operate as a power circuit breaker? _____.

ANSWERS TO REVIEW QUESTIONS

1. Must be on at the same time
2. On
3. AND
4. The output conditions for various combinations of inputs
5. SCR
6. Power circuit breaker

THE TRIAC POWER CONTROLLER

SCOPE OF STUDY

This experiment demonstrates how a triac power controller is activated by an optocoupler.

OBJECTIVES

Upon completion of this experiment, construction of circuitry, testing, and evaluation of the data, you will be able to:

1. Describe the operation of a triac control circuit
2. Suggest applications for the principles involved in the circuit
3. Test and evaluate the triac portion of a power controller
4. Communicate your understanding of unidirectional (diac) drivers to other technical and nontechnical people
5. Research optoelectronic components for controlling a high-power triac
6. Maintain and repair industrial/commercial equipment using triac couplers, given additional information and training on the overall operation of the equipment

7. Read and comprehend technical articles which describe the use of similar devices

BACKGROUND

The triac is a widely used ac power controller. It can be used for varying the voltage, such as in dimming lights, varying motor speeds, and controlling the temperature of a heater. The triac is continuously variable, and almost the full ac voltage is obtainable. High-power triacs are available for controlling voltages above 115 V ac. The optocoupler MOC3010 or MOC3011 can be used as a driver for a high-power triac.

Figure 12-1 shows the structure of a coupler used for controlling a triac. If the input voltage to the IRED of the coupler is subject to variations, the IRED should be provided with a protective circuit. If the load on the power triac is inductive and subject to in-rush surge current, snubber protective components should also be included in the circuitry.

Figure 12-2 shows a solid-state relay circuit which can be remote-controlled and used for both resistive and inductive loads. The input diode D_1 and transistor circuitry serve as protection against input voltage varia-

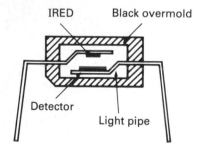

FIG. 12-1 Structure of the MOC3010/3011 coupler.

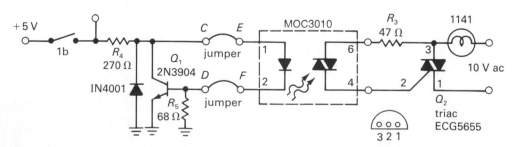

FIG. 12-2 Solid-state relay control with triac.

tions and voltage reversal. The current flow to the IRED in the coupler is also limited. The input circuitry prevents voltage variations which would seriously affect the emitter.

Because low current and low voltage are used on the input, the 5-V source can be remotely connected by means of no. 22 gage wires. Such wiring could reduce the cost of commercial installation since heavy-duty no. 14 or 12 wires are not needed. If a higher dc voltage is used for the IRED, the value of resistor R_4 can be increased. The current through the IRED should be set between 10 and 15 mA. Both R_4 and R_5 control the emitter current.

In the output section, the components provide surge protection for inductive loads, and the circuit works equally well on resistive loads. The type of output triac will depend on the voltage and current of the load. The triac shown in this experiment will handle 200 V at 6 A. At maximum rating, a heat sink would be required. In the experimental circuit, a low-voltage lamp is used to simulate a high-power load. Emphasis is placed on the input control circuitry, and not the ac power load or snubber required. In the experimental circuit, for safety reasons, a line voltage of 115 or 230 V is not used. Since these high voltages are not used, proper snubber components cannot be utilized since the required resistors would cause excessive voltage drops.

TESTS AND MEASUREMENTS

MATERIALS REQUIRED FOR EXPERIMENT

Active Devices
MOC3010 Optocoupler (triac)
2N3904 NPN transistor
1N4001 Diode
ECG Triac 5655

Others	**Resistors**
Lamp type 1141	47 Ω
Switch SPST	68 Ω
	270 Ω

Construct the experimental circuit shown in Fig.12-2. A 12-V lamp and 10-V ac line are used to simulate a higher-voltage (and power) load. The load is controlled by operation of switch 1b.

1. Connect a +5-V dc (variable) supply to switch 1b. A 12-V ac supply simulates the high voltage to the load (lamp).
2. Place 1b, the control switch, in the OFF position, and then turn on the line power. Does the load lamp light?
3. Move 1a to the ON position. Does the lamp light?
4. With the +5 V dc present, measure the voltage drop V_5 across R_5. Calculate I_5.
5. Open switch 1b and connect a 0- to 6-V variable supply to TP_1. Increase the input voltage from +5 to +6 V. A change from 5 to 6 V is a 20 percent increase. What percentage change in IRED current takes place? What percentage change in current takes place if the voltage is reduced to 4 from 5 V? What are the values of I_6 and I_4?
6. What happens to the ac lamp brightness when the input IRED voltage is varied between 0 and 5 V?
7. What is the difference between an MOC3010 and an MOC3011? (See Appendix B.)
8. What is a snubber used for?

REVIEW QUESTIONS

Test your comprehension of the subject by answering the following questions. Answer True or False on a separate piece of paper.

1. In the experimental circuit (Fig. 12-2), a −5-V supply could replace the +5-V supply.
2. As the current to the IRED is increased, the current to the output load is also increased.
3. The current rating of a triac driver is determined by the gate current of the triac.
4. The power supply for operating the IRED must be isolated from the load supply.
5. The isolation voltage is an important characteristic of the coupler.
6. The triac operates on both halves of the ac cycle.
7. High-power triacs require effective heat sinking.
8. In the experimental circuit, the IRED is current-limited by R_4 and R_5.
9. In the experimental circuit, if a voltage spike occurs, the transistor current will increase and hence the IRED current will decrease.
10. The triac can be used with only resistive loads.

ANSWERS TO REVIEW QUESTIONS

1. F
2. T
3. T
4. F
5. T
6. T
7. T
8. T
9. T
10. F

SOLID-STATE AC/DC POWER RELAYS

SCOPE OF STUDY

In this experiment, the use of zero-voltage control circuits and high-power controllers is described.

OBJECTIVES

Upon completion of the study, assignments, and evaluation of the data, you will be able to:

1. Describe the operation of a zero-voltage switching (ZVS) device
2. Suggest applications for the ZVS circuit
3. Test and evaluate circuits which utilize zero-voltage switching devices
4. Communicate your understanding of the function of a ZVS circuit to other technical and nontechnical people
5. Research optoelectronic components for use in ZVS applications
6. Maintain and repair industrial/commercial equipment using a ZVS circuit, given additional information and training on the overall operation of the equipment
7. Read and comprehend technical articles which describe the use of zero-voltage crossing devices

BACKGROUND

The electromechanical relays, both ac and dc, have control coils, isolated contacts, and specified voltage and current contact characteristics. Relays, however, are bulky, and often do not properly interface with IC logic elements or power line circuits. Solid-state relays can perform the same functions as the electromechanical devices, plus they offer some special features such as higher reliability, zero-voltage switching, direct interface with other solid-state circuits, and low cost.

Figure 13-1 shows a comparison between the sections of an electromechanical relay and a solid-state relay. The optocoupler serves as the control coil and initial contacts of the relay. The relay coil may require 6, 12, 24, or 115 V dc or ac. The LED in the coupler usually operates at 1.5 V dc, with 10 to 20 mA. The frequency of the light is either in the visible or infrared region. The series resistor enables the LED (or IRED) to operate over a wide voltage range. The LED will operate on pulses (logic) or analog signals and can directly interface with the output of other ICs, or transistors. The sensor in a coupler is isolated from the LED. Typical breakdown voltages range from 500 to 10,000 V. The isolation is analogous to the insulation between the relay, its frame, and relay contacts.

The coupler and its related circuitry can also incorporate another feature, that is, zero-voltage switching. Zero-voltage switching requires a special network. The ZVS network monitors the line voltage and uses this information to control the power contacts (on-off operation). When the coupler's IRED is interfaced with its source, consideration must be given to the type of controller used with the source. A transistor-transistor logic (TTL) circuit, if logic control is used, can sink up to 50 mA and source only 1 to 2 mA. A complementary

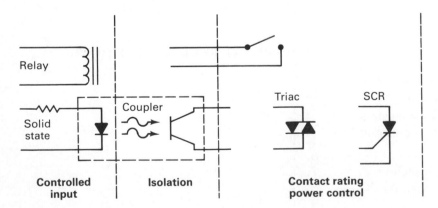

FIG. 13-1 Mechanical and solid-state relays.

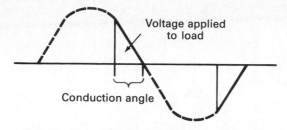

FIG. 13-2 Principle of phase control used in triac controllers.

metal-oxide semiconductor (CMOS) or metal-oxide semiconductor (MOS), however, may only be able to provide microamperes of current. Since the IRED requires currents which are typically 15 to 50 mA, additional transistor IC drivers may be necessary. When analog signals are used as the source current, Schmitt triggers or comparators are best used as drivers to the IRED. Such circuits prevent irregular lighting and noise, which are the equivalent of relay chatter.

ZERO-VOLTAGE SWITCHING

The triac power controller can operate on both the positive and negative half cycle of an ac voltage. The device makes use of phase control. Figure 13-2 shows how the sine wave is divided.

During the conduction period, the current flow is determined by the size of the load as well as the voltage of the power line source. The triac can operate up to the full line voltage while an SCR operates only up to the 50 percent level. Although the triac has many advantages, it is not without some disadvantages. The spikes produced during phase control radiate electromagnetic interference (EMI) into other equipment. Each time the triac fires, the load current, rising from zero, produces a wide spectrum of noise, and filtering at the source is often required. The sharp spikes produce voltages in the radio-frequency range. The filter circuitry, there-

fore, is referred to as a radio-frequency interference (RFI) filter.

In addition to EMI suppression, the interference can be reduced by zero-voltage (also called zero-crossing-point) switching. The device is turned on as soon as the power sine wave passes through zero. Usually this is within the range of 0 to 5 or 7 V. The technique reduces turn-on transients, which are the main source of radiation. Power to the load is more in the form of bursts of complete sine waves rather than sharp spikes. Figure 13-3 shows a typical circuit in which ZVS circuitry is used with an SCR. Transistor Q_1 controls the gate of the SCR. If the ac line voltage is below 5 to 7 V when a signal (pulse or on condition) is passing through the coupler, the SCR will be turned on since Q_1 is held off. However, if the line voltage is above the ZVS window, Q_1 conducts and the gate current to the SCR is reduced so that the SCR cannot conduct. The SCR is held off. The zener is used to provide a constant voltage to the sensor and transistor Q_1.

Figure 13-4 shows a ZVS circuit which uses a triac thyristor. The 22-Ω resistor shunts voltage changes through the bridge and prevents them from reaching the triac gate. The 100-Ω resistor limits surge and gate currents to safe operating values. The component marked by "Z" is an EMI suppressor. Essentially it is a zener diode which breaks down on transient spikes.

The rating of the triac depends on the operating voltage and load current. A load can change in value, and hence its current demand will change. A cold lamp, for example, has a high starting current (low resistance). As the wire element gets hot, its resistance increases. Also, a motor on startup has a high in-rush current. This is especially true when the motor is driving a heavy load. Figure 13-5 shows two tables indicating the in-rush currents of lamps and motors. The triac must be able to safely handle the highest currents. The motor currents are shown for the normal operating condition as well as for the locked rotor. A locked rotor, for the same motor, will increase the motor current 5 to 8 times higher than its normal operating level. In select-

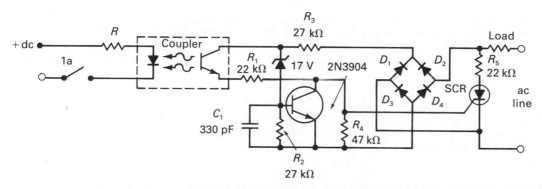

FIG. 13-3 ZVS, normally open, controller circuit.

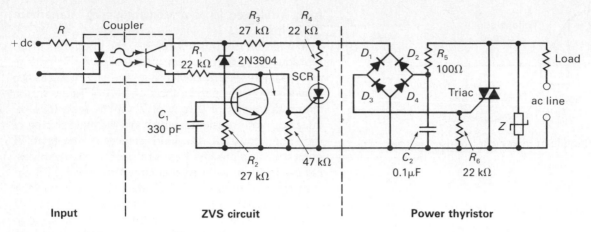

FIG. 13-4 ZVS triac controller.

ing a triac, therefore, such in-rush currents must be considered.

Figure 13-6 shows a triac circuit used in an application where high in-rush currents are not present and a ZVS circuit is not used. The coupler incorporates an SCR drive for triggering the triac. The triac acts as a normally open contact relay, with contacts rated according to the line voltage and load current. The resis-

tor R_{GK} is chosen to turn on the SCR gate, by setting the gate current, when the input voltage to the IRED is turned on.

The above relay circuits are all operated on ac power. A dc relay, with a normally open contact, exists when an SCR coupler is used. For low-level switching, such as in process instrumentation, the SCR or triac alone in a coupler may be quite adequate. The off-to-on

Wattage	Rated volts	Type	Amps steady-state rated volts	Hot/Cold resist. ratio	Theoretical peak in-rush (170 V pk) (A)	Rated (lm/W)	Heating time to 90% lm (s)	Life rated hours averg.	General Electric triac selection
6	120	Vacuum	0.050	12.4	0.88	7.4	0.04	1500	SC136
25	120	Vacuum	0.21	13.5	4.05	10.6	0.10	1000	SC136
60	120	Gas-filled	0.50	13.0	9.70	14.0	0.10	1000	SC141/240
100	120	Gas-filled	0.83	14.3	17.3	17.5	0.13	750	SC141/240
100(proj)	120	Gas-filled	0.87	15.5	19.4	19.5	0.16	50	SC141/240
200	120	Gas-filled	1.67	16.0	40.5	18.4	0.22	750	SC146/145
300	120	Gas-filled	2.50	15.8	55.0	19.2	0.27	1000	SC146/245
500	120	Gas-filled	4.17	16.4	97.0	21.0	0.38	1000	SC250/260
1000	120	Gas-filled	8.3	16.9	198.0	23.3	0.67	1000	SC250/260
1000(proj)	120	Gas-filled	8.7	18.0	221.0	28.0	0.85	50	SC250/260

For 240-V lamps, wattage may be doubled.

(a)

Horse-power	110–120 V			220–240 V			Mtr. lock-rtr. current (A)				G.E. triac' selection	
	Single-phase	Two-phase	Three-phase	Single-phase	Two-phase	Three-phase	Single-phase		Two- or three-phase		120 V	240 V
							110–120	220–240	110–120	220–240		
1/10	3.0	—	—	1.5	—	—	18.0	9.0	—	—	SC141/240	SC141/240
1/8	3.8	—	—	1.9	—	—	22.8	11.4	—	—	SC146/245	SC141/240
1/6	4.4	—	—	2.2	—	—	26.4	13.2	—	—	SC146/245	SC141/240
1/4	5.8	—	—	2.9	—	—	31.8	17.4	—	—	SC250	SC141/240
1/3	7.2	—	—	3.6	—	—	43.2	21.6	—	—	SC260	SC146/245
1/2	9.8	4.0	4.0	4.9	2.0	2.0	58.8	29.4	24	12	SC265	SC260

* Assumes overcurrent protection has been built in to limit the duration of a locked-rotor condition.
Source: Information for these charts was taken from National Electric Code, 1971 edition.

(b)

FIG. 13-5 In-rush currents of incandescent lamps and motors. *(a)* Typical current ratings. *(b)* Full-load motor-running and locked rotor currents, in amperes, corresponding to various ac horsepower ratings.

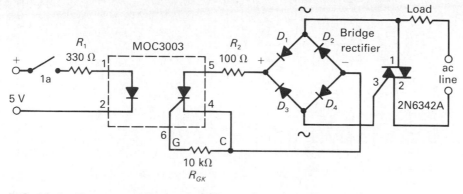

FIG. 13-6 Triac controller with SCR coupler, normally open load current.

contact resistance of the controller exceeds $10^6 \, \Omega$, and switching occurs in microseconds without such common problems as the welding of contacts or contact bounce.

The dc coupler and its circuitry are directly controllable from other ICs. The coupler has wide application with instrumentation transducers where thermocouples, thermistors, strain gages, and audio signals are incorporated. Where load currents are low, the SCR or triac in a coupler may be quite adequate by itself. If a higher current-carrying capacity or increased voltage is required, additional amplifiers can be added.

Figure 13-7 shows a circuit for a dc solid-state relay where the current-carrying capacity has been increased by the addition of transistor Q_2. Transistor Q_1 is used to increase the sensitivity of the relay. The relay contact is normally open, and its power-handling capacity depends on the ratings of Q_2. With no signal to the input of the coupler, its LED is off and the output of the sensor is high. With the base of Q_1 at a high, its collector is low, thus making the base of Q_2 low and its collector high. The output circuit is open, and current will not flow through the load resistor until the LED is turned on. By adding another inverter stage before the output power transistor, the circuit can be changed to operate as a

normally closed contact. This type of circuit is shown in Fig. 13-8.

Both ac and dc solid-state relays have been reviewed. Additional information is available from the numerous manufacturers of components. Special reference is made to the optoelectronics manuals published by General Electric and Motorola.

TESTS AND MEASUREMENTS

MATERIALS REQUIRED FOR EXPERIMENT

Active Devices
TIL111 Coupler, transistor
Q_4 Triac, 200 V ac/6 A
Q_3 SCR 6 A, 200 V
Q_2, 2N3904 Transistor, NPN
Q_1, 2N3905 Transistor, PNP
$D_1 - D_8$, 1N4001 Diodes (8) or bridges (2)
D_9 Zener, 17 V, 1N5247

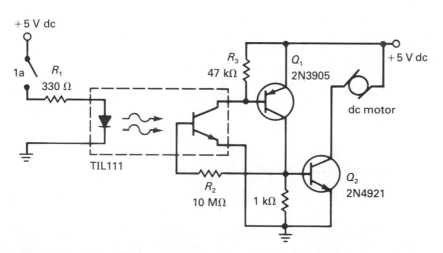

FIG. 13-7 DC solid-state relay.

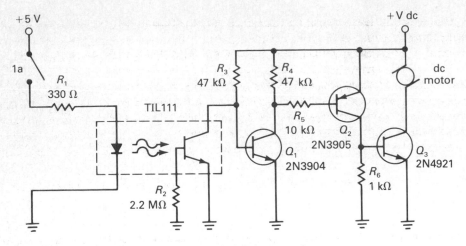

FIG. 13-8 Normally closed contact, dc solid state relay.

MATERIALS REQUIRED FOR EXPERIMENT
(Continued)

Resistors	Capacitors
27 Ω, 5 W	C_1, 330 pF
100 Ω	C_2, 0.1 μF
510 Ω (2)	
1 kΩ (3)	
10 kΩ	
22 kΩ (2)	
27 kΩ (2)	
33 kΩ	
47 kΩ (2)	
10 MΩ	

Others
100-W Lamp
120/220-V Transformer (foreign) or
   120V/120V

A ZVS circuit is presented, but it is not constructed because of potential shock hazards. If desired, the instructor can approve of its construction or can demonstrate a prewired circuit. The circuit's operation is reviewed.

Figure 13-9 shows one of the many types of ZVS circuits. The circuit is more complex than some others because the SCR Q_3 is a high-power device which requires gate current in the milliampere range. If a low-power SCR is used, Q_1 and Q_2 might not be necessary.

The triac Q_4 can be operated directly from the power line; however, a transformer provides isolation and some small margin of safety during testing. The voltage rating of the triac must exceed the line voltage.

The lamp shown is the load. Obviously, a motor could be used in its place. When the SCR Q_3 conducts, the diode bridge (D_1-D_4) acts as a switch connecting R_{14} to the triac gate. This fires the triac and reduces the

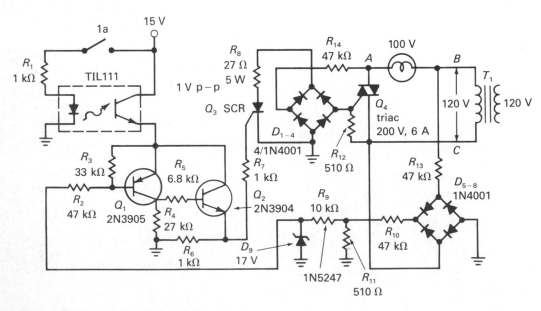

FIG. 13-9 An ac relay with ZVS control.

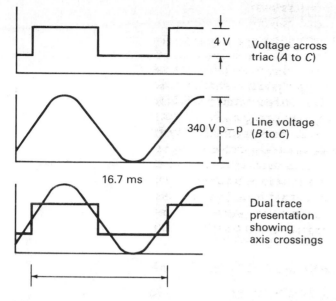

4 V — Voltage across triac (A to C)

340 V p—p — Line voltage (B to C)

16.7 ms

Dual trace presentation showing axis crossings

FIG. 13-10 Comparison of ac with ZVS.

anode voltage of the SCR causing it to stop conducting. On the following alternation, the process repeats.

The bridge rectifier, D_5-D_8, provides pulses to Q_1 for turning it on; but if the coupler is not illuminated, there is no dc voltage to Q_1 and it cannot conduct. When Q_1 conducts, Q_2 is moved to nonconduction and the SCR is turned off. In this condition, the shunting effect of the SCR is removed, and the triac conducts. A separate bridge rectifier and zener regulator circuit is used to trigger the sensing amplifier (Q_1-Q_2) since the dc output from D_1-D_4 drops to almost zero when the SCR conducts.

When point B to C of the ac input sine wave is compared to points A to C, the conduction of the triac, the waveshapes shown in Fig. 13-10 are observed.

The firing of the triac occurs after the first 5 V, and so the initial RF spikes are omitted. The ac power circuit is controlled by switch 1a. Closing 1a turns on the LED, which illuminates the phototransistor. The transistor current is the dc supply to the sense amplifier (Q_1) which is being triggered by dc pulses from the

bridge rectifier and zener. Transistor Q_2, an emitter-follower, provides the positive pulses needed for triggering the SCR.

The ac-controlled circuit is limited to the load and triac (less the triac gate) and is independent on the remainder of the circuit. The ac circuit to be controlled determines the characteristics of the triac used.

REVIEW QUESTIONS

On a separate piece of paper, complete the following statements with the appropriate word or words.

1. Could the SCR and its related bridge be used by itself as a controller? _____.
2. The ZVS controller eliminates RFI by switching during the first _____ to _____ V.
3. The electrical rating of the triac is determined by the _____.
4. The triac is a _____ switch while the SCR operates on the _____ half-cycle.
5. The solid-state circuitry described is equivalent to a(n) _____ with ac on its contacts.
6. If an electromechanical relay were used in place of the ZVS, the alternating current would flow as soon as the contactor closed. In the solid-state circuit, the first _____ to _____ of shortage is ignored.
7. The optocoupler acts as a(n) _____.
8. The optocoupler circuit is _____ of the ac-controlled circuit.

ANSWERS TO REVIEW QUESTIONS

1. Yes, the load would have a pulsating dc current.
2. 5 to 7
3. Voltage of the ac source and load current
4. Biphasic, positive
5. dc relay
6. 5 to 7 V
7. Switch
8. Independent

THE AUTOMATIC NIGHT LIGHTING CONTROLLER

SCOPE OF STUDY

This experiment demonstrates how ambient light can be used to operate remote power controllers.

OBJECTIVES

Upon completion of the study, assignments, and evaluation of the data, you will be able to:

1. Describe the operation of an ambient light controller
2. Suggest applications for this type of circuitry
3. Test and evaluate circuits which operate in a manner similar to the experimental circuit
4. Communicate your understanding of the function of a light controller to other technical and nontechnical people
5. Research optoelectronic components for light control applications
6. Maintain and repair industrial/commercial equipment using optodevices, given additional information and training on the overall operation of the equipment
7. Read and comprehend technical articles which describe the use of similar devices

BACKGROUND

Turning on and off night lights along the road is a practical application of the use of photosensors. The on-off operation of the lamp follows ambient lighting. The circuit shown in Fig. 14-1 makes use of a phototransistor sensor, a power transistor, and a lamp. The lamp could be of the high-intensity type, operating on a 6-V battery. This means that the lighting can be made portable. A variation of this circuit, which includes a triac, makes the circuit useful for high-power lighting.

In the circuit transistor, Q_2 controls the current flow through the lamp. Making the base of Q_2 more positive will reduce V_{CE}. When the transistor approaches saturation, the lamp increases in brightness (lamp voltage approaches 5 V). As the photo-Darlington sensor Q_1 approaches darkness, its collector-to-emitter voltage V_{CE} increases. As the ambient light increases, the V_{CE} of Q_1 decreases. This voltage change controls the base current of the lamp's series transistor Q_2. The simplicity of the circuit makes it very versatile. Applications include walkway lighting (as the sun goes down),

towers, street lighting, rural and remote lighting applications, and the turning on of beacons.

TESTS AND MEASUREMENTS

MATERIALS REQUIRED FOR EXPERIMENT

Active Devices
2N3904 Transistor, power, NPN
MFOD200 Phototransistor

Other	Resistors
Lamp type 1141	6.8 kΩ
	2.2 MΩ
	10 MΩ

This circuit, because of its simplicity, is easy to build. The circuit is tested to see if it follows the room's ambient light.

1. Construct the circuit as shown in Fig. 14-1. Shield the lamp with cardboard or other material, so that it does not shine across the sensor.
2. Adjust the supply voltage to $+10$ V dc before connecting it to the circuit.
3. If the room light is bright, the lamp should be off. If the room is not bright, the lamp will partially glow. What happens when the sensor is covered and it is in darkness?

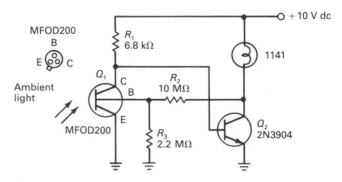

FIG. 14-1   Automatic lighting circuit.

4. Uncover the sensor, and measure V_{CE} across the sensor. What is the value of V_{CE} (Q_1) when the sensor is covered?
5. As the lamp gets brighter, what happens to the bias on the sensor (base of Q_1)?
6. The series power transistor must be capable of conducting the _____ .
7. How would you expand the circuit so that it can be used as a controller for a higher-power lamp operating from ac lines?

REVIEW QUESTIONS

On a separate piece of paper, complete the following statements with the appropriate word or words.
1. A phototransistor photosensor was used because it has a _____ gain than a photodiode.
2. The power rating of Q_2 is determined by the _____ .
3. The ratio of R_2 to R_3 determines the _____ of Q_1.
4. When the ambient light increases, the resistance of Q_1 _____ .
5. Which elements of the operating circuit determine the supply voltage?

ANSWERS TO REVIEW QUESTIONS

1. Higher
2. Lamp
3. Gain
4. Decreases
5. Lamp and Q_2

EXPERIMENT **15**

THE AUTOMATIC NIGHT FLASHER

SCOPE OF STUDY

This experiment covers the operation of a light flasher which starts to operate as darkness approaches.

OBJECTIVES

Upon completion of the study, assignments, and evaluation of the data, you will be able to:

1. Describe the operation of a light-activated flasher
2. Suggest applications for the principles involved in the circuit
3. Test and evaluate a circuit which is activated by darkness
4. Communicate your understanding of the flasher-type circuit to other technical people
5. Research optoelectronic components for application in similar circuits
6. Maintain and repair industrial/commercial equipment using flasher-controlled circuits, given the circuit and instruction
7. Read and comprehend technical articles which describe the use of light-activated sensors

BACKGROUND

The automatic night flasher provides a bright flashing light which has application as a safety marker, warning light, and entrance beacon. The portable light should have high levels of intensity and a short duty cycle. This provides maximum visibility and extended battery life. The experimental circuit is shown in Fig. 15-1.

The flasher circuit consists of a three-stage dc-coupled transistor oscillator. The sensor, a phototransistor, is used as one of the transistors. The RC time constant of R_6 and C_1 determines the frequency of oscillation or pulsation. This assumes that light is not present on the sensor. As light irradiates the sensor, its gain is reduced and the oscillation ceases. As the ambient light level gets brighter, the flashing intensity is reduced and oscillation eventually stops. With this feature, the flasher is operating at full brightness at night, when it is needed, but is extinguished during the day. The system can operate on lower-voltage batteries, such as 4.7 or 6 V (depending on the lamp). The lamp is pulsed for milliseconds at high brightness. When reinforced by a reflector, the light flashes would be observed at a distance, making it ideal for use as a safety marker.

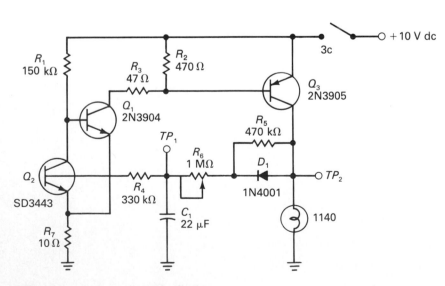

FIG. 15-1   Automatic night flasher.

TESTS AND MEASUREMENTS

MATERIALS REQUIRED FOR EXPERIMENT

Active Devices
MFOD200 Phototransistor, Q_2
2N3904 NPN transistor, Q_1
2N3905 PNP power transistor, Q_3
IN4001 Diode

Resistors	**Capacitor**
10 Ω	22 μF
47 Ω	
470 Ω	**Other**
150 kΩ	1141, 12-V lamp
330 kΩ	
470 kΩ	
1-MΩ Potentiometer	

The circuit is tested and evaluated for flashing time and switching characteristics.

1. Set the power supply voltage to 10 V, and connect this supply and ground connections to the circuit under test. The room light (ambient) should not be bright.
2. Turn on the power and cover the sensor. What is the approximate flashing rate? If the flashing does not occur, adjust R_6.
3. What happens as you uncover the sensor?
4. Use an oscilloscope to view and sketch the waveshape and amplitude of the voltage present at TP_1

and the charging current at TP_2, while the lamp is flashing.

5. Does the oscillation (flashing) depend on the light's output which is irradiating the sensor (covering the sensor)?
6. The pulse current through the lamp could be higher than the normal operating lamp current, true or false?
7. How could the brightness be further increased?
8. Simulate sunrise by increasing the light shining on the sensor. What happens to the brightness of the lamp?

REVIEW QUESTIONS

On a separate piece of paper, complete the following statements with the appropriate word or words.

1. The flashing rate is determined by a(n) _____ time constant.
2. The power rating of Q_3 is determined by _____ .
3. The brighter the ambient light, the _____ the flasher.
4. The waveshape of the oscillator is a _____ wave.
5. Resistor R_4 controls the _____ of the phototransistor.

ANSWERS TO REVIEW QUESTIONS

1. *RC*
2. The lamp current
3. Dimmer
4. Square
5. Pulse amplitude to the sensor

POWER CONTROL WITH LOGIC GATES

SCOPE OF STUDY

This experiment describes how power-controlling circuits can be activated by remote-control logic devices.

OBJECTIVES

Upon completion of the study, assignments, and evaluation of the data, you will be able to:

1. Describe the operation of a logic-controlled circuit
2. Suggest applications of the principles involved in this circuit
3. Test and evaluate basic logic-switched power circuits
4. Communicate your understanding of the use of logic components for power control to other technical and nontechnical people
5. Research optoelectronic components for other similar applications
6. Maintain and repair industrial/commercial equipment which uses logic devices for power control, providing additional information and training are available on the overall operation of the equipment
7. Read and comprehend technical articles which describe the use of logic-controlled devices

BACKGROUND

In all the previous circuits, the LED or IRED was controlled by a mechanical switch. In this experiment, a logic gate is used to control the emitting diode. The most common logic gate family is the transistor-transistor logic (TTL), the 54/74 series. These ICs provide sufficient current (source or sink) to interface with couplers. The MOS or CMOS devices require additional buffer amplifiers since the logic gates operate in the microampere range. Some manufacturers, such as General Electric, have designed their couplers (H74 series) to match with logic elements.

Figure 16-1 shows a two-input positive NAND gate (7400) controlling a phototransistor coupler. Figure 16-2 shows the gate controlling an SCR coupler. The SCR can control a peak forward voltage of 200 V (400 V for the H74C2) and an rms forward current of 0.3 A with a peak dc forward current of 10 A. The NAND gate requires that both inputs be a 1 (2 to 5 V) for the output to be a 0 (LED conducts when the gate

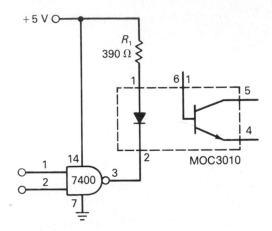

FIG. 16-1 NAND gate control of coupler.

output goes to zero). For the output to be a 0, the input voltage is typically 0.22 to 0.4 V. When the gate output goes to zero, it can sink 16 mA of current. The short-circuited output current of the gate, however, is 55 mA. If both inputs to the NAND gate are tied together, they will both go high or low at the same time. The gate acts as an inverter (7404).

The MOS and CMOS can drive the coupler if a high-gain (h_{fe}) transistor is added as a buffer. Figure 16-3 shows a CMOS gate and buffer connected to a coupler. The CMOS can operate down to 3 V; therefore, the value of R_1 should be chosen so that both the IRED and the CMOS can operate from the same dc supply. Here Q_1 is a high-gain transistor. The D3854 (General Electric), for example, has an h_{fe} of 400 to 3000 and an I_C of 100 mA (continuous).

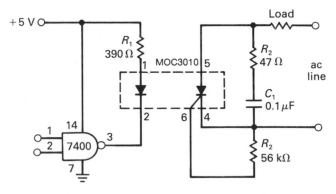

FIG. 16-2 NAND gate control of SCR coupler.

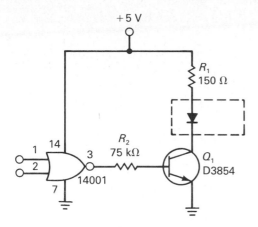

FIG. 16-3   CMOS to coupler driver.

A logic element, shown in Fig. 16-4, can be used in place of a mechanical switch to control a LED or IRED in a coupler. The 7400 series of TTL, with its higher driving current, can directly control the coupler, whereas the CMOS ICs with a lower output current will require a buffer amplifier.

TESTS AND MEASUREMENTS

MATERIALS REQUIRED FOR EXPERIMENT

Active Devices
MOC3010
SN7400 NAND gate
5655 ECG triac

Resistors	**Other**
47 Ω	Switch SPST
330 Ω	1141, 12-V lamp
470 Ω	

A triac coupler is wired so that it can be controlled by a logic gate. A low-voltage ac supply and lamp are used to simulate a high-voltage line and load current.

PROCEDURE

1. Construct the circuit shown in Fig. 16-4. Use an ac line voltage of 10 V. A fixed 5 V is used for the IRED supply. Temporarily leave out the connection between pin 4 of the coupler and pin 2 of the triac (gate). (Open switch 2b.)

2. Place an ohmmeter between pins 4 and 6 of the coupler. When the IRED is not turned on (switch 1b is open), what is the resistance, in ohms? This is the no-light resistance (or a 0 logic level).

3. Close switch 1b to turn on the +5 V and illuminate the IRED (a logic 1). What is the resistance between pins 4 and 6 of the coupler?

4. Close switch 2b. Turn on the power by closing switch 1b. Does the load light go on?

5. Measure and record the ac voltage across the load lamp.

6. Open and close switch 1b. Does the logic gate act as a switch?

7. Is the logic element used to source or sink the IRED current?

8. If the current through the IRED were increased by reducing the resistance value of R_6 from 330 to 270 Ω, would the voltage across the load lamp be increased?

9. Is the diac element in the coupler fully on when the IRED is on?

10. How could you vary the intensity of the load lamp?

11. If terminal 2 of the coupler went to ground through a switch, would it perform the same as the logic element?

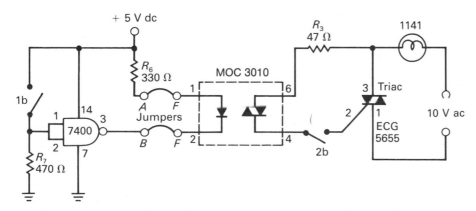

FIG. 16-4   Logic-controlled triac.

REVIEW QUESTIONS

On a separate piece of paper, complete the following statements with the appropriate word or words.

1. A logic gate can operate the same as a(n) _____ .

2. When both inputs to a NAND gate go high, its output goes _____ .

3. An SN7400 (TTL) has a (higher, lower) current output than a CMOS.

4. In a coupler the sensor has a _____ isolation from the emitter.

5. The size of the triac is determined by the _____ .

ANSWERS TO REVIEW QUESTIONS

1. Switch
2. Low
3. Higher
4. High-voltage
5. Load

PART FOUR

APPLICATIONS TO DIGITAL COMMUNICATIONS

SCOPE OF STUDY

Data communications or digital control is the transfer of information, in a pulse sequence format, by electronic or optoelectronic means. Terminals communicate with computers, control devices cause operational changes in machines, and human voices, digitalized, glow through light pipes by the thousands. In this section, basic digital transmitters and receivers are investigated.

OBJECTIVES

Upon completion of this study, assignments, and evaluation of the data, you will be able to:
1. Describe the manner, in general, in which terminals are able to communicate with computers, microprocessors, and logic-controlled machines
2. Explain the operation of each circuit tested
3. Suggest applications of the principles involved in each section
4. Communicate your understanding of the subject matter to the layperson or engineers
5. Maintain or repair equipment which uses similar types of circuits
6. Read and comprehend technical articles which describe the use of similar devices and circuits

BACKGROUND

Data communication is generally considered as a system of communication between a computer and its input-output (I/O) terminals. In the broader sense, data communication principles are also applicable to the control of machines, rockets, and communications of the human voice. *Data communication* is the transfer of information in a digitalized format. Until recently, such data transfer was performed mainly by wire cables. Optoelectronics now offers an alternative which is lower in cost, rugged, and immune to electromagnetic interference (EMI).

Some general concepts and terms relating to data communications are introduced. Various methods of communication call for different types of couplers, circuits, and hardware. In the latter part of the discussion, attention is directed to light transmitters and receivers.

Although artificial voice translators are currently being introduced, the main method of communicating is still by means of keyboard terminals and video displays. To make use of such terminals, information must be coded into a format which the equipment can understand. Humans use alphabetical characters to formulate words; in computers, data is translated into digital (binary) pulses, which are formed into specific codes. Each digital pulse is called a *bit*, and bits are grouped to form codes. Some typical codes currently in use are Baudot, Numeric Control, Flexwriter, ASCII, and EBCDIC.

DATA TRANSMISSION CODES

The Baudot code, used in telex communications, is made up of 5 binary digits per character. Character readers, in part, are rated according to their printing speed. The term *baud* is interchangeable with cycle per second, or hertz. One baud is one bit per second. A printer capable of printing 30 characters per second has a speed of 30 Hz.

The American Standard Code for Information Interchange (ASCII) is a seven-level code allowing for a 128-character set, as compared with 32 characters per set used in the Baudot system. The ASCII system is used in one of two versions, namely, the full or limited ASCII:

	Full ASCII	Limited ASCII
Codes	128	97
Printable characters	95	64
Nonprintable, control codes	33	33
Upper- and lowercase alphabet, numerals, punctuation, symbols		Uppercase only

While the Baudot character contains 5 bits, the ASCII uses up to 7 bits per character. A character is usually a language unit of information presented in a bit sequence code.

SYNCHRONOUS AND ASYNCHRONOUS TRANSMISSIONS

Synchronous data transmission is the flow of data in one direction at a time. The procedure is not well suited to remote-keyboard-operated terminals. Asynchronous transmission is better suited to two-way communication systems. Each character being transmitted has a start pulse, followed by 7 bits of data, and then one or more stop pulses. The start pulse is usually a 1, while the stop contains two pulses. While 7 bits contains the character code, the other pulses (three) provide synchronization. As previously indicated, ASCII uses a total character count of 11 bits. In ASCII, the 7-bit code is followed by a *parity* pulse. This parity pulse is transmitted in one of four forms, namely, *even, odd, mark,* and *space.* The pulse verifies that the character is being properly received. It provides character verification. In even parity, the total number of 1s transmitted adds to an even number, while in odd parity it adds to an odd number. In mark parity, the mark is a 1; and in the space parity code, the space is a 0. For systems to be compatible, they must be able to transmit and receive the same code.

BUFFERS

Buffers are storage devices which accumulate data for transmittal to a printer. Printers are mechanical devices, and data flow may exceed the printer speed. The buffer, therefore, can store the data until the printer can catch up. The buffer is a short-term memory which prevents data loss. The buffer operates on the first-in, first-out (FIFO) principle. The first data to enter the buffer will be the first out to the printer.

WIRED COUPLERS AND MODEMS

Terminals have been connected to their computers via hard-wire lines or through communications links. A third option, fiber-optic links, enables transmission to take place by means of a digital pulsed light beam connected through an optical conductor.

When hard-wire lines are used, digital pulses cannot be transmitted directly over the lines. The pulses must be converted to tones (analog signals). This conversion is performed by a *modem* (a *mo*dulator/*dem*odulator) which is coupled to the telephone by an acoustical or other coupler. At the terminal the pulses form tone bursts, and at the computer the tones are again demodulated to pulses. Directly wired modems are more effective than acoustical coupling devices. The losses are less as are the chances of error.

In a fiber link the pulses are converted to light modulations by a transmitter. At the receiver end the light is demodulated, and the original pulses are produced. The optical system is similar to a directly wired modem, but may be more efficient.

SIMPLEX AND DUPLEX LINKS

Both telephone wired lines and fiber-optic lines use simplex, duplex, and half-duplex links. In a simplex link, communications go in only one direction whereas in a duplex system communications can go in both directions *at the same time.* In a half-duplex system, communications go in either direction on the same fiber, but in only one direction at a time. One side is transmitting, and the opposite side is receiving. In a fiber-optic link, only one fiber cable is required for a simplex and half-duplex. A full duplex may require the use of two fiber cables and dual connectors.

DATA ENCODING

Several forms of encoding data are used in information transfer. In one coding system, NRZ (*non*return to *zero*), a string of 1s is encoded as a continuous high-level pulse. When the pulse state goes low, the level also goes low and stays low until there is a pulse state change. Figure A shows the waveshape produced by an NRZ encoding system. The system stays either high or low.

In a data transmission receiver, which may be ac-capacitor-coupled, long durations of high-level pulses will cause some data to be lost. In addition, in both codes, a clock signal is not transmitted. In a third technique, called the *Manchester encoding system,* a synchronizing clock pulse is transmitted. A typical pulse train is shown in Fig. B.

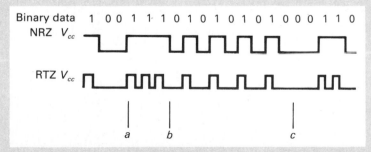

FIG. A   NRZ and RTZ encoding systems.

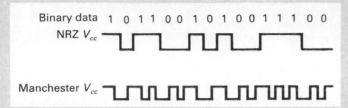

FIG. B   NRZ and Manchester data encoding.

In the Manchester encoding scheme, the level changes with each data pulse. The large number of transitions enables the system to encode clock pulses. Even if a string of 1s or 0s is being transmitted, the clock rate is still available. Another problem with the NRZ and return-to-zero (RTZ) systems is that both have a tendency to shift and maintain a receiver's automatic gain control (AGC) at a high level, thus moving the amplifiers toward saturation. The Manchester scheme does not affect the gain, but it requires a still higher bandwidth.

Another method of pulse encoding is the *bipolar* method. This is compared with NRZ in Fig. C. In this system, the LED or IRED stays at a quiescent level until a pulse change is received. Zero pulses turn the light source on to a full level for a short duration while 1s reduce the current level for short durations.

The advantages of the pulse bipolar system over the RTZ encoding system is that the transmitter always sends data at a fixed pulse width so that it places no restrictions on the input signal other than its maximum frequency. Since the light never turns off, the receiver's AGC system is not shocked, but rather maintains a smoother reference level.

In Experiment 20, a bipolar transmitter/receiver is investigated. In the initial laboratory investigations, some basic transmitter modulators will be investigated as well as some receivers.

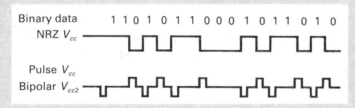

FIG. C   Pulse bipolar encoding.

DIGITAL TRANSMITTER/RECEIVER CIRCUIT

SCOPE OF STUDY

This experiment introduces some basic circuits for converting digital pulses to light pulses for transmission on fiber-optic cables.

OBJECTIVES

Upon completion of the study, assignments, and evaluation of the data, you will be able to:

1. Use TTL-compatible components and light emitters to develop transmitter circuits
2. Determine the maximum pulse repetition frequency of a transmitter
3. Use Schmitt trigger detectors for converting light impulses to logic pulses

BACKGROUND

Digital chips containing inverters, NAND, NOR, and Schmitt triggers are typical of the components which can be used to drive LEDs. If the chip cannot sink or source the required emitter current, a transistor can be used to provide the additional emitter current.

The circuit shown in Fig. 17-1 is a digital light transmitter in which the chip can directly sink the LED current. By using the same circuit but changing the logic of the chip to a 75453 (OR), the truth table becomes:

**Truth Table for
75453 (OR)**

A	B	C
L	L	0
L	H	1
H	L	1
H	H	1

Number 1

A 1 at the output (C) enables the LED to radiate its light while a 0 indicates that the LED is dark. Resistor R_1 is adjusted in value to produce the required LED current. A 0.5-W resistor may be required for higher currents. The pulse width and pulse repetition frequency are limited by the LED. Figure 17-2 shows a similar circuit which can be used as a pulse transmitter, and it is TTL-compatible.

The truth tables for two IC chips are shown in Table 17-1. Again, the value of R_1 determines the LED current. The value of the V_{cc} bypass capacitor is determined by the repetition frequency. It may vary in value from a 0.01 to a 1-μF capacitor.

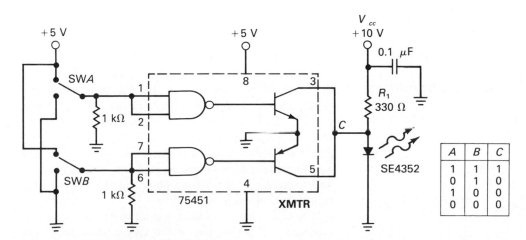

FIG. 17-1   TTL digital transmitter.

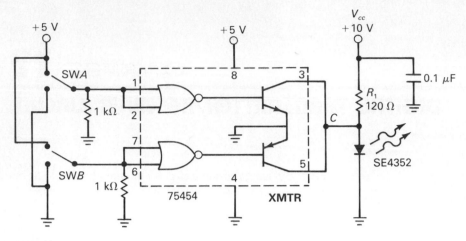

FIG. 17-2   Pulse transmitter.

TABLE 17-1
Truth Table for 75452 and 75454

75452 (NAND)			75454 (NOR)		
A	**B**	**C**	**A**	**B**	**C**
L	L	1	L	L	1
L	H	1	L	H	0
H	L	1	H	L	0
H	H	0	H	H	0

Figure 17-3 shows a transmitter which consists of a TTL interface and a LED driver. As previously indicated, the TTL element must interface with a common digital family, such as TTL. The LED driver also provides the forward current required by the LED. In addition, the driver must have a switching speed which is compatible with the incoming baud rate. If CMOS logic is used in the interface, the driver converts voltage level changes to current level changes. The LED, mounted in a holder or socket, must provide light or near-infrared into a port which is properly aligned with a fiber-optic cable. In the overall design, consideration must be given to:

1. The ease in adjusting the LED current
2. Isolation of the LED current from line power changes
3. Isolation of the LED current from temperature changes
4. Minimizing current pulses in the V_{cc} lines, which also supplies other circuits

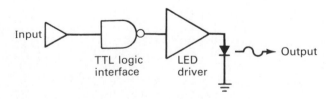

FIG. 17-3   TTL interface and LED driver.

THE RECEIVER

The block diagram of a typical optical receiver is shown in Fig. 17-4. The first stage converts the photodiode current change to a voltage change, which is then amplified by the linear amplifier. Voltage transitions are used to trigger a comparator or Schmitt trigger circuit which directly interfaces with other TTL circuits. The optical detector, which serves as a current source, receives pulses of optical energy emanating from the end of the fiber-optic cable. The small current that flows, passes through a resistive load which is paralleled by component and circuit capacitance.

The greater the value of this capacitance, the lower the rise time and frequency response of the receiver. The circuit has a high input impedance and is subject to noise, EMI, and other disturbances, which are easily coupled into the voltage amplifier. The differentiator circuit eliminates the trailing edge of the pulse which is not needed for the triggering of the logic element.

A number of semiconductor manufacturers have produced a detector which combines many of the above receiver designs in a single unit. The Honeywell SD4324, shown in Fig. 17-5, is one such device which contains a Schmitt trigger and TTL driver. The internal 10-kΩ resistor can be externally paralleled in order to increase the rise time.

At low data rates, under 10 kilobits per second, the distortion is minimal; but as the data rate increases, pulse stretching (T_s) will occur. Figure 17-6 shows a graph of the received pulse power versus pulse stretching. The frequency is a 50 percent duty cycle operating at 50 kHz. Stretching at the positive rise ($+T_s$) of the output pulse indicates sufficient power at the receiver, and stretching at the low-going portion ($-T_s$) indicates that the received power is low.

The LED current at the transmitter can be increased or decreased depending on the type and length of the fiber cable. The duty cycle also affects both the power output of the LED and the degree of pulse stretching.

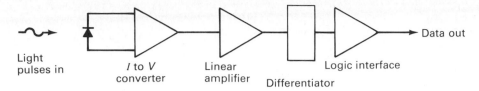

Light
pulses in I to V Linear   Differentiator   Logic interface
            converter   amplifier   → Data out

FIG. 17-4   Optical receiver.

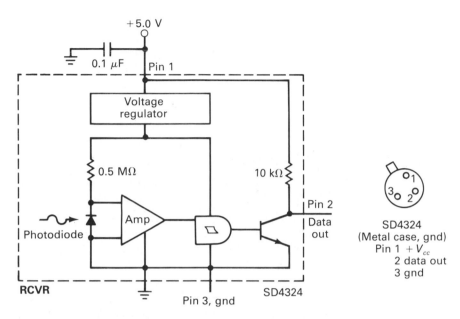

FIG. 17-5   Complete monolithic receiver with Schmitt trigger.

Figure 17-7 shows the electrical data for the SD4324, whose wavelength is highly sensitive, between 700 and 900 nm. The receiver, depending on the data to be transferred, can be a simple interfacing device, as shown, or a complex system. The design of the transmitter, the fiber-optic link, and the receiver can be expressed in one word — *transparent*. The entire system between data in and data out should be transparent — as though it did not exist. The engineer need only know that what goes in will come out — undistorted.

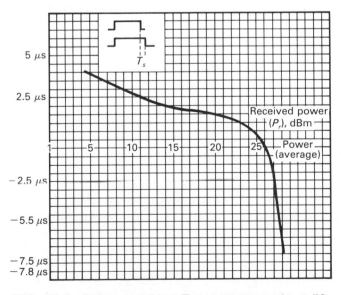

FIG. 17-6   Pulse stretching T_s in microseconds at 50 kHz.

TESTS AND MEASUREMENTS

MATERIALS REQUIRED FOR EXPERIMENT

Active Devices
MC75451 IC peripheral driver, AND
SE4324 Schmitt trigger, photodetector/
 connector (228043-1, A)
SD4352 IRED, emitter/connector (228043-1, A)

Resistors   **Capacitors**
1 kΩ (2)   0.1 μF (2)
330 Ω

Others
SPDT switch (2)
Fiber cable, 50 cm/connectors (A 228041-1
and 22840-1, 2 each)

Parameter	Test condition	Symbol	Min.	Typical	Max.	Units
Input sensitivity	$\lambda_p = 820$ nm	P_{min}		2	5	μW
Field of view	Note 1	F_oV		40		Degrees
High-level logic output voltage	$P_{in} = 10\ \mu$W $V_{cc} = 5$ V $I_{ol} \geq 100\ \mu$A	V_{oh}	2.4			V
Low-level logic output voltage	$P_{in} < 0.5\ \mu$W $V_{cc} = 5$ V $I_{ol} = -16$ mA	V_{ol}			0.4	V
Logic output Propagation delay Time	$V_{cc} = 5$ V dc $R_L = 390\ \Omega$					
Low to high		t_d (L-H)		3.0		μs
High to low		t_d (H-L)		3.0		μs
Logic output Transition time	$V_{cc} = 5$ V dc $R_L = 390\ \Omega$					
Low to high		t_r		80		ns
High to low		t_f		10		ns
Output low Supply current	$V_{cc} = 5$-V input $V_{cc} = 16$ V Power $= 0$	I_{ccL}	4 5	6 7	12 15	mA mA
Output high Supply current	$V_{cc} = 5$-V input $V_{cc} = 16$ V Power $= 5\ \mu$W	I_{ccH}	2 3	4 5	10 12	mA mA

Note 1: Angle between 50% response points. Fifty percent response points are defined as the angle at which twice the on-axis irradiance is required to trigger the detector.

FIG. 17-7 SD4324 technical data for Schmitt receiver: electrooptical characteristics ($T_c = 25\degree$C unless otherwise specified).

In the laboratory activities, you will evaluate the logic of a transmitter and the characteristics of a Schmitt trigger detector.

1. Construct the circuit shown in Fig. 17-8, and evaluate its truth table. Use wire jumpers in place of the switches.
2. Ground A (a 0), and switch the B input between 0 and $+5$ V. Record the output (from C to ground). Test the remainder of the truth table, and determine whether your findings confirm the statements shown.
3. Measure the voltage drop across the IRED feed resistor, and calculate the IRED current I_f.
4. Connect switch A to a high. Switch B should be in an open position so that a function generator, set to a 1-kHz square wave at a 5-V peak positive pulse, can be connected. Observe and record V_o. View also, with a second oscilloscope channel, the

pulse voltage across the IRED. Does a high input pulse produce a high-going output?

5. Shift the duty cycle to 25 percent. Increase the frequency, and observe the output for changes in the rise time and pulse width. Determine the approximate frequency range where these features are observed.

Receiver Tests

6. Connect the supply voltage to the Schmitt detector which is housed in a fiber cable connector. Connect a 50-cm fiber cable between an output and the Schmitt trigger detector's input connector. Set the generator to a pulse rate of 5 to 10 pulses per second. Observe the outputs at both the IRED and the Schmitt trigger. Are the two pulses of the same phase?

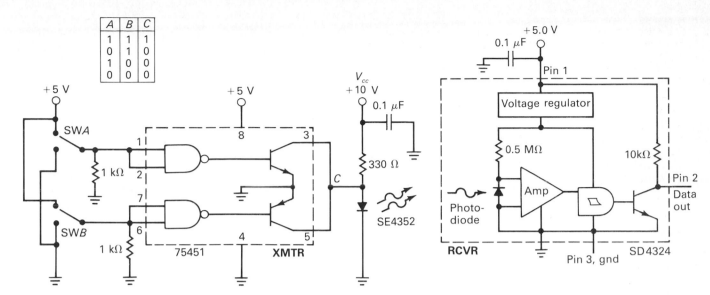

FIG. 17-8 Experimental circuit.

7. Determine again at what frequency pulse stretching occurs. This time view the Schmitt trigger output as compared to the output of the IRED. Where does T_s occur?

8. What is the pulse width of the Schmitt trigger detector when the generator is set to 1 kHz and the positive input from the generator is 50 μs? If the fiber cable is disconnected, does the output fall to zero?

9. What determines the current flow I_f of the IRED?

On a separate piece of paper, answer the following questions True or False.

10. Fiber cables do not produce pulse stretching.

11. Too high a light level, or too low a level, could change the pulse width from the Schmitt trigger detector.

12. If a logic element cannot source or sink the emitter current, a CMOS driver is required.

REVIEW QUESTIONS

On a separate piece of paper, complete the following statements with the appropriate word or words.

1. I/O terminals are in the form of _____.

2. A *bit* refers to a digital _____.

3. The baudot code is used in _____ communications.

4. A parity pulse verifies that characters are being properly _____.

5. A modem is used to convert a(n) _____ signal to _____ tones.

6. Simplex means communications in _____ direction.

7. In NRZ coding, a string of 1s causes a continuous _____ series of pulses.

8. A receiver is more easily synchronized when _____ encoding is used.

9. When a transparent link is used, the user is not aware of the _____.

10. The output of the SD4324 is a(n) _____.

ANSWERS TO REVIEW QUESTIONS

1. Keyboards, video displays, printers
2. Pulse
3. Telex
4. Received
5. Analog, pulse
6. One
7. High-level
8. Bipolar or Manchester
9. Presence of the link
10. Pulse

ONE-MEGABIT NRZ TRANSMITTER

SCOPE OF STUDY

The experimental tests involve the use of an inverter chip and a transistor driver to form a basic NRZ transmitter, with a data transmission rate of 10 megabits.

OBJECTIVES

Upon completion of the study and evaluation of the data, you will be able to:

1. Describe how data pulses are converted to light pulses for transmission on fiber cables
2. Test and evaluate digital optoelectronics transmitters, when provided with circuit diagrams and test data
3. Measure the power output of a digital transmitter by use of a light-power instrument

BACKGROUND

In a typical digital system, the coding format used might be NRZ (nonreturn to zero). In this format, a string of 1s is encoded as a continuous high level. Only when there is a change of state to a 0 does the signal level drop to 0. In RTZ (return-to-zero) encoding, the first half of a clock cycle is high for a 1 and low for a 0. The second half is low in either case.*

Figure 18-1 shows an NRZ and RTZ waveform for a .binary datastream. Note that between *a* and *b* the RTZ pulse repetition rate is at its highest. The highest bit rate requirement for an RTZ system is a string of 1s.

---

* The discussion on the data encoding format is provided through the courtesy of Motorola.

The highest bit rate for an NRZ system is alternating 1s and 0s, as shown from *b* to *c*. Note that the highest NRZ bit rate is half the highest RTZ bit rate, or an RTZ system would require twice the bandwidth of an NRZ system for the same data rate.

To minimize drift in a receiver when NRZ encoding is used and a long string of 1s is transmitted, an alternate scheme is needed. In addition to drift prevention, the ac receiver coupling may result in lost data. An alternate to using NRZ is RTZ. With RTZ data flow, data is not lost in ac coupling since only a string of 0s results in a constant signal level; but that level is itself 0. In both the NRZ and RTZ methods, for any continuous string of either 1s or 0s, the receiver might lock up. An alternate method of coding is therefore suggested.

Another format, called Manchester encoding, solves the receiver lock-up problem. In Manchester, the polarity reverses once each bit period regardless of the data. This is shown in Fig. 18-2. The large number of level transitions enables the receiver to derive a clock signal along with the data. Even if all 1s or all 0s are being received, the receiver also obtains a clocked synchronizing pulse chain.

In many cases clock recovery may not be required. It might appear that RTZ would be a good encoding scheme for these applications. However, many receivers include automatic gain control (AGC). During a long stream of 0s, the AGC could crank the receiver gain up; and when 1s begin to appear, the receiver may saturate. A good encoding scheme for these applications is pulse bipolar encoding. This is shown in Fig. 18-3. The transmitter runs at a quiescent level and is turned on harder for the short duration of a data 0 and is turned off for the short duration of a data 1.

For additional details on encoding schemes, see recent texts on data communications or pulse code modulation.

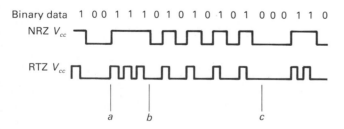

FIG. 18-1   NRZ and RTZ encoding systems.

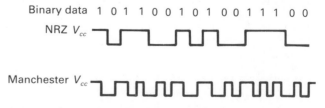

FIG. 18-2   Manchester data encoding.

Binary data   1 1 0 1 0 1 1 0 0 0 1 0 1 1 0 1 0

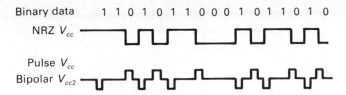

FIG. 18-3   **Pulse bipolar encoding.**

TESTS AND MEASUREMENTS

MATERIALS REQUIRED FOR EXPERIMENT

Active Devices	Resistors
SE4352 LED, emitter	47 Ω
2N3905 PNP transistor	510 Ω
SN7404 IC, inverter	1 kΩ
	10 kΩ
Other	
Diode IN4148 (3)	**Capacitor**
	100 pF

In this experiment you will test and evaluate a high-frequency NRZ pulse transmitter. The basic transmitter circuit could be used in a microprocessor data link. Figure 18-4 shows the experimental circuit.

The IRED is turned on when Q_1 is turned on, and $D_2 + D_3$ is used to ensure the turnoff of the LED. Here D_1 prevents the reverse biasing of Q_1 (base-emitter breakdown). The value of resistor R_4 is determined from

$$R_4 = + \frac{V_{cc} - 3.0V}{I_f}$$

where V_{cc} is the supply voltage and I_f is the desired IRED current. For a 10-V supply and $I_f = 0.05$ A, $R_4 = 140\ \Omega$. A nominal value of 150 Ω could be used. For a +5-V supply, $R_4 = 40\ \Omega$. The 5-V supply would make the transmitter compatible with other TTL circuits; hence it is used in the laboratory work.

1. Set the power supply to +5 V. Close switches 3a and 4a. Connect the circuit shown in Fig. 18-4. Measure the IRED current I_f by recording the voltage drop across the 47-Ω resistor.
2. With no input data pulses, measure and record the voltage to ground at the test points shown in the circuit [that is, find U_1 input (TP_1), output (TP_2), Q_1, base, emitter, D_2, D_3, and TP_3 (measured to ground)].
3. Set your function generator to produce positive pulses of 3 to 4 V in amplitude and 2 to 3 μs wide when $F = 100$ kHz. Connect the generator to the data input (TP_1) terminal of the transmitter. View this input and at the same time view TP_3. Does the output follow the input?
4. Does a zero input cause the output to fall? Does it drop to zero?

The following questions relate to the background discussion.

REVIEW QUESTIONS

On a separate piece of paper, complete the following statements with the appropriate word or words.
1. Does a bipolar pulse output enable the receiver's AGC to return to zero? _____ .
2. A modem is generally used for data transmission _____ lines.
3. Modems are used to change logic pulses to _____ .

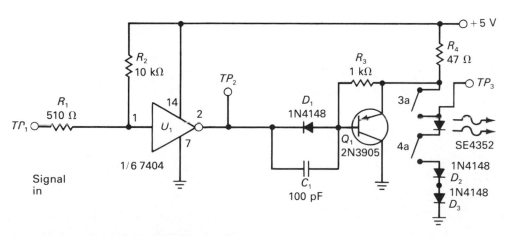

FIG. 18-4   **One-megabit NRZ transmitter.**

4. A Baudot rate of 150 is equal to _____ bits per second.
5. A character is made up of logic _____ .
6. In ASCII a character is equal to _____ bits.
7. In ASCII, of the 10 bits, _____ defines the _____ .
8. Between a terminal and a computer, _____ transmission is best used.
9. The SE4352 operates in the _____ range.

Additional Study

Use a light-to-power converter to measure the power output from 1 m of fiber-optic cable when the transmitter's pulse input is a 50 percent duty cycle (at 1 kHz) and when the duty cycle is approximately 10 percent. Observe what happens to the peak power when the duty cycle is increased.

ANSWERS TO REVIEW QUESTIONS

1. Yes
2. Via telephone
3. Tone signals
4. 150
5. Bits
6. 10
7. 7 bits, character
8. Asynchronous
9. Infrared

A DATA COMMUNICATIONS RECEIVER

SCOPE OF STUDY

Data communications receivers should have a wide bandwidth, in order to be able to handle a large dynamic range, and be TTL-compatible with other logic circuits. In this experiment, a receiver and transmitter are used as a communications system.

OBJECTIVES

Upon completion of the study unit, laboratory tests, and evaluation of the data, you will be able to:

1. Test and evaluate a data communications receiver, when provided with appropriate test instruments, circuits, and technical data
2. Determine by physical measurements the bandwidth of the receiver

3. Evaluate the transmitter and receiver sections of a communications link

BACKGROUND

One type of optoreceiver input transducer is the photodiode. It produces minute currents that must be amplified and appear as voltage levels compatible with logic devices. Its low sensitivity is offset by its high speed of response (1 to 10 ns), which makes it an ideal choice for high-speed data communications. Figure 19-1 shows a circuit diagram of the basic section for a receiver.

The transmitter/receiver system, coupled by a fiber-optic cable, can be used to interface telecommunications, microprocessor, and control systems. In the laboratory experimental work, a receiver will be linked to a transmitter by a fiber-optic cable, and the total system evaluated.

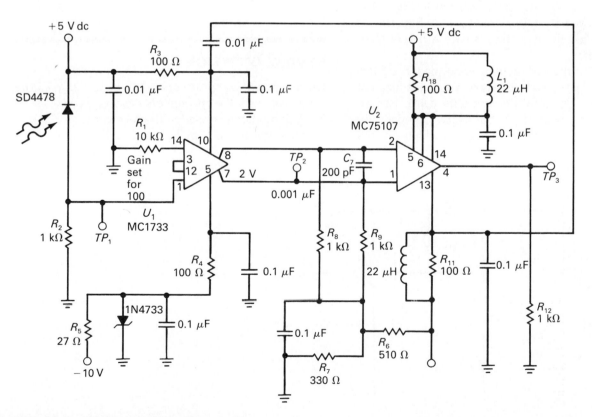

FIG. 19-1   Data communications receiver.

TESTS AND MEASUREMENTS

MATERIALS REQUIRED FOR EXPERIMENT

Active Devices
SD3443 Photodiode
MC1733 High-gain, wideband amplifier
MC75107 High-gain line receiver
Transmitter from Experiment 18

Resistors	Capacitors
100 Ω (4)	0.1 μF (6)
510 Ω	0.01 μF (2)
270 Ω	200 pF
330 Ω	
1 kΩ (4)	

Others
IN4733 5-V zener diode
22 μH (2) RF choke coils
50 cm Fiber-optic cable with components

A data transmitter and receiver are linked by a fiber-optic cable, and the system is tested by using a square wave and a pulsed signal.

1. Construct or connect a prewired PC board which contains the receiver circuit shown in Fig. 19-1. Connect the transmitter circuit used in Experiment 18, and link the two circuits with a 50-cm fiber-optic cable.
2. Initially test the system by feeding the transmitter with a 3-V p-p, 1-kHz square wave from a function generator. Use a 50 percent duty cycle. Record the waveshape at TP_1 of the IRED. Compare the rise time of the signal source to that which appears at TP_1 (across the 1-kΩ resistor). What is the signal source?
3. Record the waveshape that appears at the receiver output, TP_2.
4. Compare the rise time of the receiver output with that of the function generator.
5. The rise time is a function of bandwidth, and bandwidth (BW) can be determined by the equation $BW = 0.36/t$, where BW is in megahertz and the rise time t is in microseconds. What is the bandwidth of this system?
6. To faithfully reproduce a square wave, a system should have a bandwidth that is 10 times greater than the fundamental frequency which is to be reproduced. What, then, is the highest-frequency square wave that this system is capable of reproducing with good fidelity?
7. Are there any lower limits to this system's bandwidth? Why or why not?
8. Change the input signal to 1 MHz, and reduce the duty cycle so that a positive pulse width of 0.2 μs appears. Increase the duty cycle to 50 percent. Does the system reproduce both signals faithfully?
9. Measure the signal that appears across the photodiode load resistor R_1. How much current flows through the photodiode? How much power is present?
10. Is the output signal of sufficient amplitude to drive a TTL device?

REVIEW QUESTIONS

On a separate piece of paper, complete the following statements with the appropriate word or words.
1. A photodiode is used in the receiver rather than a phototransistor. Why? _____ .

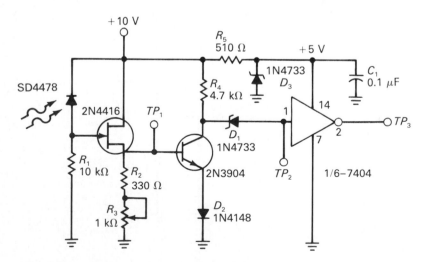

FIG. 19-2 Alternate experimental circuit.

2. The gain of the receiver is controlled by the resistance between terminals _____ and _____ .
3. Is the bandwidth of the MC1733 IC greater than 3 MHz? _____ .
4. The value of resistor R_1 controls the gain and _____ of the receiver.
5. The MC75107 produces an output whose waveshape is a(n) _____ .
6. Resistors R_5 and R_7 determine the _____ .
7. What is the lowest frequency the receiver can handle? _____ .
8. Can the MC1733 handle both analog and digital signals? _____ .

The circuit in Fig. 19-2 can be used for additional experimentation or as an alternative to Fig. 19-1. The bandwidth and sensitivity of the circuit in Fig. 19-2 are not as good as those in Fig. 19-1, but are adequate for the experiment.

ANSWERS TO REVIEW QUESTIONS

1. Its wider frequency response and greater speed
2. 3, 12
3. Yes
4. Bandwidth
5. Square or rectangle
6. Switching voltage of the MC75107
7. dc since it is dc-coupled
8. Yes

BIPOLAR TRANSMISSION

SCOPE OF STUDY

The bipolar pulse operation of an IRED in the transmitter of a communications link enables the system to encode three operating states, namely, a logic 1, a logic 0, and a median state. The IRED is never turned off, as in the NRZ and RTZ operations. Data is always transmitted at a fixed pulse rate, and the AGC circuitry in the receiver always maintains a reference level.

OBJECTIVES

Upon completion of the study and by answering the required questions, you will be able to:

1. Describe how a typical system design, which includes a transmitter and receiver, is used to transmit pulse data
2. Explain how a transmitter can be made to transmit logic 1s, 0s, and a synchronizing frequency
3. Use engineering application notes, as provided by a manufacturer, for system designing

BACKGROUND

This chapter, unlike most others in this book, does not contain a laboratory experiment. You are provided with application notes, prepared by David Stevenson of Motorola, and you will use these notes for studying a communications system. The notes were originally titled "MFOLOZ, Theory and Application." Although this type of IRED emitter can be used, other types are also applicable. The emphasis is on the operation of a transmitter and receiver circuit. The layout design of the PC boards has been omitted, and you should refer to the *Optoelectronic Device Data Manual* by Motorola for additional data. The design of a fiber-optic communications link should be such that the entire linkage appears to be "transparent" to the user. In other words, the designer who wishes to take advantage of some of the benefits of fiber-optic digital data transmission need not know any more about the system's modules than that they take in TTL and give out TTL.

The original application of the link was for systems requiring only direct current to 200 kilobits and point-to-point system lengths of up to 1000 m. The specific modules described or other similar ones can be easily designed. For the more ambitious student, the circuits can be constructed and the system evaluated. Those individuals who read the description and review the questions will also benefit from the broader concepts and understanding of such designs.

Before the circuit analysis is begun, the general specifications of the modules should be highlighted. First, both the transmitter and the receiver circuits were designed for single 5-V power supply operations. As previously stated, the bandwidth capability is direct current to 200 kilobits, and depending on the particular optical fiber that is used, the transmission path can be extended up to 1000 m.

Physically, both module housings are identical, being approximately $2 \times 2 \times 45$ in. The module base is configured similar to a large in-line package having eight pins fixed in two rows of four each. Spacing between the pins is 0.400 in, and spacing between the two rows is 1.670 in. Optical input and output ports are provided using AMP Optimate fiber connectors. The modules are designed with removable covers so that the PC boards and associated components can be accessed even when the circuits are in operation.

TRANSMITTER

Circuit analysis will begin with the transmitter shown in Fig. 20-1. The basic requirement of this circuit is to convert TTL voltage levels to corresponding current pulses through the light-emitting diode. The original circuit used an MFOE102F; however, an MFOE71 or SE4352 can be used. Furthermore, the transmitter provides for tertiary or pulse bipolar encoding format. Basically with the pulse bipolar encoding format, the LED operates in three distinct states. During idle modes in data transmission, the LED drive assumes a median level which is midway between logic 1 and logic 0. During positive-going transitions (logic 1 to logic 0), the LED is momentarily driven at approximately twice the median or quiescent level. The advantage of the pulse bipolar format over the standard binary RTZ format is that the transmitter always sends data at a fixed pulse width, so it places no restrictions on the input signal other than maximum frequency. Another advantage of this type of transmission is that during idle modes of data transmission the light source is not turned off; so if the receiver incorporates automatic gain control, it always maintains a reference level.

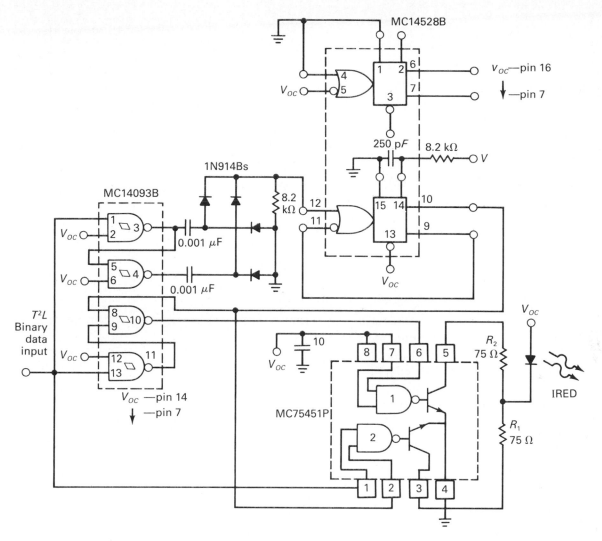

FIG. 20-1 Transmitter circuit.

Beginning at the transmitter input (Fig. 20-1) the binary TTL signal drives the input of a two-input NAND Schmitt trigger (¼ MC14093). This gate forms an inverter by virtue of its second input being tied to V_{cc}. This inverted signal is then split, and part of it is input to pin 5 of the second NAND Schmitt trigger. The result is that the signal at pin 4 is essentially the input waveform, and the signal at pin 3 is its complement. These two complementary signals are differentiated by 0.001-μF capacitors and rectified by a full-wave bridge formed by the four 1N914B diodes. The result is that for every transition of the input, either 0 to 1 or 1 to 0, a positive pulse is applied to the "set" input of the MC14528B monostable multivibrator. The MC14528B multivibrator is programmable so that the output pulse width can be determined by an external RC time constant at pin 14. The values chosen give a pulse width of approximately 2 μs which is adequate for 200-kilobit transmission. This, then, will be the pulse width of the current pulses applied to the LED to represent logic 0 and logic

1 transmission. Notice that the MC14528B is actually a dual monostable, only half of which is used.

The remaining two NAND Schmitt triggers are used to gate the proper timing pulses to the MC75451P dual NAND input peripheral driver. The operation of this device is such that when the transmitter is in its idle mode, that is, the current through the LED is at the median level, the current path in this state is as shown in Fig. 20-2.

The value of idle current flowing through the LED is a function of V_{cc} and R_1 and can be calculated by

$$I_{\text{idle}} = V_{cc} - V_f - \frac{V_{\text{sat}}}{R_1}$$

where V_f is the forward voltage drop of the LED, V_{sat} is the on state voltage of the MC75451, $V_{cc} = 5$ V, $R_1 = 75$ Ω, and $V_f = 1.2$ V. The idle current is approximately 50 mA.

To understand the other two states of the pulse bipolar transmitter, it is necessary to evaluate the sig-

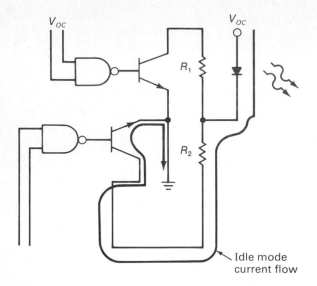

FIG. 20-2 Idle-mode current flow.

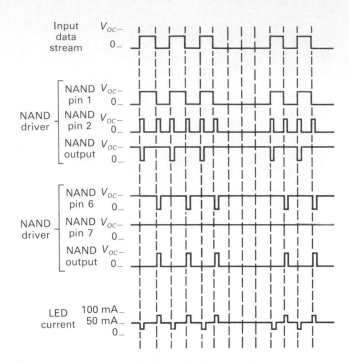

FIG. 20-3 Circuit waveforms.

nals present at the inputs to both NAND drivers at each transition point of the input datastream. The waveform at pin 1 of NAND driver 1 is that of the input data. The waveform at pin 2 is the 2-μs pulse produced by the monostable multivibrator. Before the waveform at pin 6 can be derived, it is necessary to evaluate the action of the other two NAND Schmitt trigger gates. The input waveform is buffered and inverted by NAND driver 4 (input pin 13, output pin 11). This inverted waveform is NANDed with the 2-μs pulse output of the monostable, and the result is a 2-μs negative pulse at each negative transition of the input (1 to 0). This signal at NAND driver 3 pin 10 is connected to pin 6 of NAND driver 2. Since pin 7 is held at V_{cc}, this results in the output of the NAND gate going high (logic 1) for 2 μs at every negative transition of the input waveform. The resulting output of both NAND drivers is shown with respect to the input waveform in Fig. 20-3. For every positive transition of the input both NAND gate outputs are low, meaning the LED is turned off for 2 μs. For each negative transition of the input, both NAND outputs are high; and since R_2 is equal to R_1, the LED is driven at twice the median current level for 2 μs. At all other times, the LED is driven at the median level.

RECEIVER

The entire receiver is constructed from two CMOS integrated circuits. The MC14573C is a quad operational amplifier, and the MC14574C is a quad comparator.

The detector used for this receiver is a high-speed photodiode. This detector can be thought of as a current source whose output current is proportional to the input optical flux or light level. The receiver output device is a voltage comparator, so that between the two

some kind of current-to-voltage conversion and amplification must take place. The current-to-voltage conversion takes place at U_1 (Fig. 20-4). The theoretical gain of this amplifier, which is fixed by the 1-MΩ feedback resistor, is 1 V/μA. This, in turn, is followed by amplifier U_2 whose gain is fixed at 20 by the 5.1-kΩ input resistor and the 100-kΩ feedback resistor. The integrating amplifier formed by U_3 clamps the output reference level of U_2 to a voltage fixed by the values of R_1 and R_2. In this case these are both 5.1 kΩ, so the reference voltage is half of V_{cc}, or 2.5 V. Also U_3 tends to cancel voltage offsets produced by U_2 by feeding the signal back to the U_2 input. This allows the receiver to be dc-coupled, and the component count and cost are reduced.

The output of U_2 is then fed to comparator U_5, which provides additional amplification, and it boosts the signal to TTL levels. Comparator U_6 is used to improve hysteresis and invert the signal, so that the output waveform is in phase with the original datastream applied to the transmitter. Finally, the 2.5-V reference voltage is buffered by U_4 to prevent transients, produced by the comparators, from interfering with the front-end amplifiers and reducing the need for additional filtering.

REVIEW QUESTIONS

On a separate piece of paper, complete the following statements with the appropriate word or words.

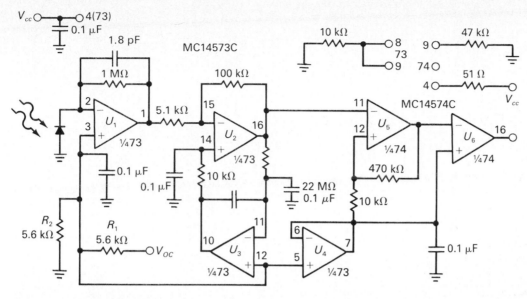

FIG. 20-4   Receiver circuit.

1. The bipolar system of data communications enables what three forms of data to be transmitted? _____ , _____ , _____ .

2. In the bipolar system, the LED is _____ .

3. The transmitter converts voltage pulses to corresponding _____ pulses via light transmission of the LED.

4. When data is not being transmitted, the LED is operating at a _____ level.

5. When data is being transmitted, the pulses are of a(n) _____ pulse width.

6. In an NRZ system, the transmission of 1s is encoded as a continuous _____ level. When the zero is transmitted, the carrier returns to _____ .

7. In an RTZ system, the receiver's AGC system could be _____ when a string of 1s appears after a string of 0s.

8. An RTZ system requires a bandwidth _____ as great as the NRZ system for the same data rate.

9. Limiting factors to bandpass, and hence data rate, are caused by the _____ , _____ , _____ , _____ , and _____ .

10. In the system described, the pulses transmitted have a pulse width of _____ μs.

11. The output from the receiver must be _____ compatible with other digital circuitry.

12. A system can pass a 2-MHz square wave with a 50 percent duty cycle. How many megabits could it pass? _____ .

13. In the receiver circuit, U_1 is a(n) _____-to-voltage converter.

14. The gain of U_1 is determined by the _____ .

15. In a transparent data link, the output pulse from the receiver is of the same phase as the _____ to the transmitter.

ANSWERS TO REVIEW QUESTIONS

1. 1s, 0s, and a reference frequency
2. Never turned off
3. Current
4. Median
5. Fixed
6. High, zero
7. Saturated
8. Twice
9. LED, detector, fiber cable, stray capacitance, response of amplifiers
10. 2
11. TTL
12. 4 megabits
13. Current
14. 1-MΩ feedback resistor
15. Input pulse

PART FIVE

FIBER-OPTIC CABLES

There are two kinds of fiber-optic cables: one type has a plastic core, and the other has a glass core. Plastic cables have a high level of attenuation and are usually operated within the light wave frequency range. Glass cables with very low losses are operated in the near-infrared and infrared regions. Part 5 focuses on the losses that occur in fiber-optic cables and the methods by which to compute these losses. Measurements are made of various lengths of cables; the photometer is used to evaluate the light transmission capability of sample cables.

Part 5 includes information about the attachment of terminations to fiber-optic cables so as to make short lengths. When fiber-optic cables are interfaced with a light transmitter and a light receiver, the systems designer must take into consideration all losses between the transmitter and receiver. This process is known as *flux budgeting design* and is studied in Experiment 22. In the telephone industry, fiber-optic cables are rapidly replacing metallic cables. Telephone cables require a repeater station approximately every mile, whereas the new, low-loss glass cables require a repeater station only every 70 to 120 miles.

In 1984, AT&T started laying two fiber-optic cables capable of transmitting 144 video channels. These cables, laid under the Atlantic Ocean, link a base station in New Jersey with base stations in London and Paris. In spite of the new satellite communications systems, fiber optics is competing for the handling of large quantities of high-speed traffic.

APPLICATIONS OF FIBER-OPTIC CABLES

SCOPE OF STUDY

In the past, the two main methods of conducting electric energy were via wire or magnetic fields (wireless). Electric power transmission is highly dependent on copper and aluminum cabling, while telecommunications and data transmission can utilize either wire or wireless systems.

The transmission of information or power control via plastic or glass optical cables offers the designer-engineer a third option. Although flashlights or signal lamps have been used for sending Morse code or flashing signals of events, they are inadequate for modern communications. In this experiment the basic concepts of fiber-optic cables and their hardware are covered. Light-emitting devices and sensors used in this study have been described in previous experiments.

OBJECTIVES

Upon completion of this study unit, which includes laboratory testing and the evaluation of materials and circuits, you will be able to:

1. Describe some of the advantages and disadvantages of using optical-coupled systems which use fiber cables
2. Describe some of the types of cables currently being produced
3. Describe how light travels through a fiber-optic cable
4. Measure the losses in some fiber-optic cables
5. Use connectors for joining cables to emitters and detectors
6. Set up a fiber-optic link for communicating between a transmitter and receiver

BACKGROUND

Media writers have been using words such as *LEDs, lasers, fiber optics, light beams, sensors,* and *magnetic fields* to describe new product features, science fiction programs, and equipment in the arsenals of war. Most of these terms relate to the rapidly emerging technology of optoelectronics. Fiber-optic cables play a major role in this technology. In this experiment a few concepts are developed for the practical use and application of fiber-optic materials in telecommunications.

The German-born Swiss physicist Albert Einstein verified what another German physicist, Max Planck, described as radiation being in the form of discrete units called *quanta.* Einstein proposed that light travels through space in a quantum form, which he termed *photons* — particles of energy. These particles could travel through a vacuum, space, gases, water, glass, or plastic. Glass and plastic are the main elements used in the making of fiber-optic cables, hence the material conducts photons.

ADVANTAGES OF FIBER OPTICS

Photons have no electric fields that can be disturbed by electrical machinery, lightning, high-voltage fields, or other disturbances so common to wire and wireless transmission. Fiber-optic cables can withstand much more abuse than a copper wire of equal size. In addition, fiber-optic cables weigh far less than copper, have a size advantage, can withstand all kinds of weather, and are immune to most liquids.

As a minimum, 50,000 voice signals can be transmitted on a 1-mm fiber cable while a copper cable composed of 900 pairs of no. 22 gage wire can handle only 10,000 voices. The following is a summary of some of the advantages of using fiber-optic cables:

1. Electromagnetic interference (EMI) and cross-talk susceptibility are greatly reduced.
2. The cables are liquid and nuclear-radiation-retardant.
3. Cables are lightweight and small.
4. Optic cables have a wider bandwidth for the same physical size.
5. Fibers are not hazardous in combustible areas.
6. Longer cable runs between repeaters are feasible.
7. High-temperature operation is not a major problem.
8. Links are secured against signal leakage.
9. Cables are mechanically strong.

Figure 21-1 compares the features of three types of cables. Fiber cables, on the negative side, are not easily tapped, such as for making a splice. The fittings of components to the cables have not been standardized. Connectors, such as produced by AMP Inc., may fit some Honeywell components but may not fit components by another manufacturer.

Property	Twisted pair	Coaxial	Fiber optics
1. RFI/EMI/noise immunity	No	No	Yes
2. Total electrical isolation	No	No	Yes
3. High transmission security	No	No	Yes
4. No cross-talk	No	No	Yes
5. Low cross-talk	No	Yes	Yes
6. No spark/fire hazards	No	No	Yes
7. No short-circuit loading	No	No	Yes
8. No ringing/echoes	No	No	Yes
9. Temperatures to 300°C	Yes	Yes	Yes
10. Temperatures to 1000°C	No	No	Yes
11. 200-MHz Bandwidth for 300 m	No	No	Yes
12. EMP immunity	No	No	Yes
13. Low cost	Yes	No	Yes

FIG. 21-1 Comparison chart for communication cables. (*Courtesy of Galileo Electro-Optics Corp.*)

In general, fiber-optic cables offer far more advantages than wire for communications. Hence cables are destined to be in greater demand, and optodevices will become another major growth technology.

TYPES OF FIBER OPTICS

Fibers can be classified by their core material. Glass is an efficient material for smaller, high-performance cables. Larger and less efficient cables have a plastic core. These cables, however, are more rugged.

Fiber cables are made in a variety of sizes and configurations. There are single conductor fibers from 0.005 mm (5 μm) to 1.5 mm, and multiple conductor cables with over 60 cores are available. Typically, practical core sizes vary in diameter from 0.25 to 1.0 mm. The outer coating of the fiber may be made of polyvinyl chloride or polyurethane, and generally the jacket is tightly applied to the fiber. In multi-core cables, strength-bearing steel cores are also included. Fibers come in three types: single-mode step-index, multimode step-index, and multimode graded-index fibers. The cable selection also depends on whether the light source is a laser or LED. The LED generally produces incoherent light in the red and infrared regions, and such light sources are commonly used in the simpler multimode step-index fibers. Fiber sizes of 5 to 1000 μm make up the multimode fibers.

Figure 21-2 shows the structure of the three types of cables. The first coating around the core is called the *cladding,* and a final wrapping provides protection from the outside elements. The single-mode step-index cable is smaller, generally 2 to 10 μm, propagates in only one mode, is highly efficient, and is suitable for high-speed

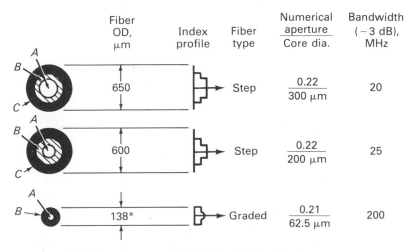

	Fiber OD, μm	Index profile	Fiber type	Numerical aperture / Core dia.	Bandwidth (−3 dB), MHz
	650		Step	0.22 / 300 μm	20
	600		Step	0.22 / 200 μm	25
	138*		Graded	0.21 / 62.5 μm	200

*With lacquer removed, the fiber OD is 125 μm.

A—fiber core   B—cladding   C—plastic coating

FIG. 21-2 The forms of fiber cable structures.

transmission. Because of its small size, it is physically difficult to work with. Its high-speed characteristic makes it suitable for data and video transmission.

The step-index fiber derives its name from the reflective index design. Light travels through air or a vacuum at a faster speed than through plastic or glass. The transmission of light through a fiber occurs at 60 to 70 percent of its speed through air. The transmission through fiber is in the range of 2×10^8 m/s. In terms of time, this is approximately 5 ns/km.

The multimode graded-index cable has a multilayer core in the form of concentric rings, each having a lower refractive index. Reflections from the cladding are far fewer, and the light follows a somewhat sinusoidal path with dispersion being under 8 to 10 ns/km.

LIGHT TRANSMISSION

Figure 21-3 shows how light is reflected along the cable with some reflections coming from the cladding as well as from the fiber core. Since the fiber's core and the cladding have different indices of refraction, the light reflects from the walls of the core. Some light, however, passes through the core and is reflected from the cladding, with some energy being lost in the cladding. Some cables exhibit a lower loss in the infrared region, while others have a lower attenuation in the visible wavelength.

Figure 21-4 shows a wavelength attenuation curve for an ESKA cable. As can be seen, the minimum loss area is in the wavelength range of 580 to 600 nm. Obviously, if this cable is used with an infrared emitter and detector, greater losses will exist. For short-length runs, the additional losses are not critical since a higher emitter output and increased detector sensitivity can compensate for the difference.

Several terms are used to describe a cable's characteristics. As indicated, the core size is important, as are the indices of reflection and numerical aperture (N.A.). The smaller the index ratio between the cladding and the core, the less the dispersion, the higher the bandwidth, and the lower the power losses.

NUMERICAL APERTURE

The *numerical aperture* defines the light-gathering capability of the fiber and the angles at which light is

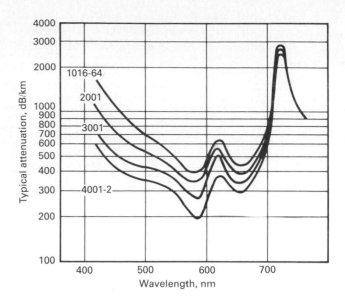

FIG. 21-4   ESKA cable, SH series.

reflected from the plastic (or glass) walls. The smaller the N.A. value, the higher the cable's bandwidth, the lower the power losses, and the smaller the cable. See Fig. 21-5.

Cables with N.A. values of 0.1 to 0.2 are usually high-frequency cables used for long runs. Short-run cables have N.A. values of 0.3 to 0.5. The larger N.A.-rated cable has greater dispersion and hence causes distortion of pulses and other signals being transmitted. With cable dispersion, pulse edges start to slope, and the pulse frequency must be lowered to prevent pulses from overlapping.

Each fiber has an N.A. rating, which is generally between 0.1 and 0.6. This rating corresponds to the light entry into the cable between half-cone angles of 9° and 33°. As indicated, fibers having a broader bandwidth and lower attenuation typically exhibit smaller N.A. values. One manufactured cable, the ESKA model SH-4001, has an N.A. value of 0.5. This cable is suitable for short runs and for applications requiring lower frequencies.

Step-index fibers are made of a quartz (silica) core with silicone cladding, whereas graded-index fibers

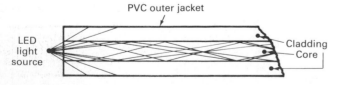

FIG. 21-3   The zigzag path of LED light in a multimode step-index fiber.

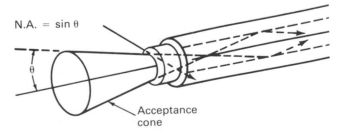

FIG. 21-5   The acceptance cone is related to the numerical aperture, which defines the light-gathering ability of the fiber.

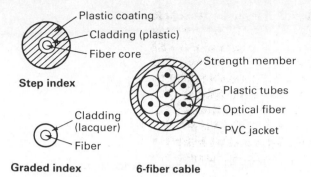

Step index

Graded index　　**6-fiber cable**

FIG. 21-6　Fiber cables — construction.

have a core of silica, cladding, and an additional coating. The graded-index fibers have a broader bandwidth (200 MHz compared to 20 to 25 MHz for step-index fibers). For shorter runs, however, the step-index cable is adequate.

Figure 21-6 shows the construction of the two types of fibers as well as a multifiber cable. The graded-index fiber is smaller in diameter, has a broader bandwidth, but may also have alignment problems. Graded-index cables, used in long runs, are generally more costly than step-index cables.

DISPERSION

Dispersion, typically measured in nanoseconds per kilometer, relates to the spreading of a light pulse as it travels along the fiber. A pulse seen at the output is wider than the input pulse. Dispersion limits a fiber's bandwidth. Pulse rates must be slow enough that dispersion does not cause adjacent pulses to overlap; i.e., the detector must be able to distinguish between them.

Consider the pulse trains in Fig. 21-7 showing how dispersion limits frequency. Two trains of pulses, one faster than the other, are injected into the same fiber. In both cases, the pulses are spread by dispersion. For the slower train, the interval between pulses is sufficient to allow each pulse to be distinguished. In the faster train, however, the interval is so short that individual pulses merge into one long, indistinguishable pulse. Dispersion, which can be related to pulse rise and fall times as in copper cable, therefore is the limiting factor in determining a fiber's bandwidth.

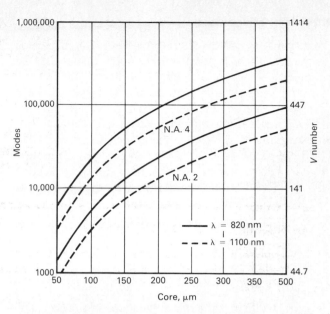

FIG. 21-8　The *V* number and number of modes are determined by fiber characteristics. Mode and *V* number values are approximate.

Two types of dispersion are modal, arising from the different paths of light in various modes, and material, arising from the different velocities of different wavelengths. In a single-mode fiber, which exhibits no modal dispersion, material dispersion is the sole frequency-limiting mechanism.

The approximate number of modes for a graded-index fiber is

$$N_m = \frac{V^2}{4}$$

Figure 21-8 shows how differences in fiber parameters can affect the number of modes. A fiber with a 50-μm core, for instance, can support more than 1000 modes. When the *V* number is about 2 (more precisely, 2.405), the fiber supports a single mode, and there is no modal dispersion.

Dispersion in a short length of fiber is proportional to length. If dispersion is 5 ns for 100 m, it is 10 ns for 200 m. For long lengths where mode coupling is present, however, dispersion becomes proportional to the square root of fiber length.

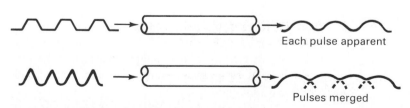

FIG. 21-7　Dispersion limits signal frequency.

87

Material Dispersion

Material (also called spectral or intramodal) dispersion is generally less significant than modal dispersion, except, of course, for a single-mode fiber. The spectral width of the source determines material dispersion and ranges from 2 nm for a laser diode to over 40 nm for a LED. The peak wavelength also affects dispersion. A LED operating at 820 nm and having a spectral width of 38 nm will result in dispersion of 3 to 3.5 ns/km. Material dispersion in the same fiber can be reduced to 0.2 to 0.3 ns/km by using a laser diode operating at 1140 nm and having a 3-nm spectral width.

Material dispersion can be approximated by

$$D_m = \frac{\lambda(\Delta\lambda)}{c} \frac{d^2n}{d\lambda^2} L$$

where λ is the average or peak wavelength, $\Delta\lambda$ is the spectral width, n is the refractive index of the core, and L is the fiber length.

Manufacturers typically combine modal and material dispersion and specify a single dispersion figure. Often a figure of merit of the bandwidth-distance product, usually given in megahertz-kilometers, replaces the dispersion figure. The product of the signal frequency and the distance that the signal is sent through the fiber must not exceed the figure of merit for good signal quality. For example, a fiber with a bandwidth-distance product of 200 MHz·km allows a signal of 200 MHz for 1 km, 400 MHz for 0.5 km, or 100 MHz for 2 km. The figure of merit relates to dispersion: a low-frequency signal can travel a longer distance than a high-frequency signal before dispersion makes the signal quality unacceptable. The bandwidth-distance product is a reasonable approximation of fiber performance if not stretched to extremes. A 200-MHz·km fiber might not permit 20 GHz for 10 m or 200 GHz for 1 m.

The advantages of low dispersion in a fiber are accompanied by tradeoffs against other fiber characteristics. Low N.A. or small core diameter offers low dispersion, but it is difficult to couple a small fiber to sources and other fibers. Dispersion may be controlled by careful selection of a low-N.A., narrow-spectral-width source such as a laser.

Modal Dispersion

Modal (or intermodal) dispersion in a step-index fiber can be approximated by simple geometry. The path length for a meridional ray is

$$P(\theta) = \frac{L}{\cos\theta}$$

where θ is the angle between the ray and the fiber axis (see Fig. 21-5) and L is the axial length of the fiber. The velocity of a ray in the core is

$$v = \frac{c}{n_1}$$

The delay time between an axial ray and any given meridional ray becomes

$$D_s = \frac{P(\theta) - L}{v}$$

When θ is equal to the acceptance cone half-angle, maximum dispersion is approximated.

The delay difference between an axial ray and a meridional ray is also approximated by

$$D_s = \frac{nL}{c} \frac{1}{\cos\theta^{-1}}$$

The maximum delay is seldom reached, however, because high-order modes are more sharply attenuated than low-order modes and because mode coupling transfers energy between modes. Such coupling occurs at bends in the fiber and at points of variation in the core's composition and diameter. Mode coupling reduces the mean spread between the highest- and lowest-order modes. In addition, dispersion in long lengths of fibers is not strictly proportional to length.

The V number, or normalized frequency, is a fiber parameter that takes into account the core diameter, the wavelength propagated, and the N.A.:

$$V = \frac{2\pi d}{\lambda} \text{ (N.A.)}$$

From the V number, the number of modes in a step-index fiber can be approximated:

$$N_m = \frac{V^2}{2}$$

BANDWIDTH

Bandwidth is another important characteristic of a fiber cable. Usually specifications are given in terms of megahertz per kilometer. A cable rated at 80 MHz/km may exhibit an attenuation of -3 dB at 80 MHz at the end of 1 km of cable.

Although fibers are available in many sizes, generally the smaller-size cables with lower aperture numbers have the widest bandwidth.

The bandwidth is affected by the dispersion of light rays from the fiber's walls. Again, light wave bouncing relates to the angle at which light enters the fiber (the numerical aperture). The angling rays arrive at the end of the fiber at different times, and this affects the bandwidth of the cable.

The SH4001 has an 80-MHz response for 50 m of cable. The limited bandwidth in some cables causes attenuation, pulse stretching, and the distortion of pulse edges. When pulse stretching occurs, two pulses in a series may overlap and appear at the output as a double-hump single pulse. The distortion of a pulse's rise time results from both material and model dispersion. Both are related to flux transmission. For short-length runs, dispersion of the pulse may not be a problem; however, in long runs, dispersion limits the rate of data transmission.

ATTENUATION

Besides pulse stretching and distortion, the cable causes an attenuation of the signal's power. Cables are specified in terms of their loss, such as X dB/km. For example, a cable may be rated at 10 dB/km at 820 mm. This means that at the end of 1 km (3280 ft) the output pulse still contains 10 percent of its original power. The attenuation, in decibels, is expressed by

$$\text{Attenuation, dB} = 10 \log \frac{W_{\text{in}}}{W_{\text{out}}}$$

The wavelength should also be stated since a cable may have a minimal loss in the visible range and a higher loss in the infrared range.

Optical fibers are available with attenuation ratings which range from 0.1 to 10,000 dB/km. Light levels are measured in terms of power, and a photometer is used for actual measurements. The light level may be expressed by using the symbol "dBm." This references the power level to 1 mW.

Suppose 500 μW is -3 dBm (since $1000/500 = 2$ and this is equal to 3 dB) when light energy enters the fiber. At the end of 1 km, 10 dB of loss has taken place. The power leaving the cable is $\frac{1}{10}$ (10 dB is a power ratio of 10/1), or 50 μW. This equates to -13 dBm. As a mathematical expression,

$$\text{Power, dBm} = 10 \log \frac{PW}{1 \text{ mW}} \qquad 0 \text{ dBm} = 1 \text{ mW}$$

where PW equals power, in watts.

The loss in a cable is often due to fiber impurities such as hydroxide (OH) ions, which are trapped in the glass, or plastic, during the cooling process. Fortunately, some cables have their minimum loss at a wavelength of 800 to 900 mm where some LEDs and detectors are the most sensitive.

LINKS IN FIBER-OPTIC CABLES

The cable is part of an overall communications system. On long-line cables, repeaters are used to reconstruct the waveshape and spacing of the pulses. Figure 21-9 shows a simple link where pulse-code modulation (PCM) is used in telephone communications.

The PCM converter changes analog signals to pulse-code-modulated signals which are multiplexed. Through the fiber flow rectangular pulses which follow the μ-255 telephone communications system used in North America or the equivalent CCITT A-Law system adopted by foreign countries. The converter, also called the CODEC (coder-decoder), feeds an amplifier and LED with time-shared pulses. The repeater receives distorted pulses which are used to trigger a pulse generator whose pulses are similar to those originally generated. The fiber link transmits the pulse chain to receiver stations where they are decoded back to analog signals. Such systems, using wideband hair-thin fibers of silica, one of the most abundant materials on earth, could replace thousands of tons of copper wire. With the broad bandwidth of thin fibers, 50,000 voice channels or 30 TV channels per fiber are feasible. Other applications include the following:

Cable television
Data systems between terminals and computers
Telecommunication systems
Instrumentation and process controls
Power control and power transfer

INTERCHANGEABLE COMPONENTS

One of the current problems is the standardization of connectors, component sizes, and shapes of devices which interface with the fiber cables (see Fig. 21-10). Alignment of the LED with the fiber depends on the effectiveness of the connector. A LED with or without a lens will focus at different locations on the cable ending, hence will affect the total internal reflections within the cable. Some fiber cable connectors are made to fit the TO-18 size component (such as some used by Honey-

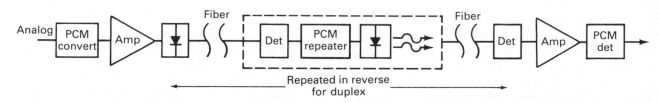

FIG. 21-9 Simplex telephone line, using PCM.

	Single wire	Twisted pair	Solid dielectric coaxial cable	Foamed airspace coaxial cable	Flat flexible cable	CATV grade coaxial	Fiber optic
Physical size	Small	Small	Small	Small	Medium	Large	Small
Mechanical integrity	Very good	Very good	Very good	Good	Excellent	Excellent	Fair
Weight	Very low	Very low	Low	Low	Moderate	Heavy	Low
Cross-talk immunity	Poor	Fair	Very good	Very good	Good	Very good	Excellent
Bandwidth	Very limited	Very limited	Limited	Limited	Very limited	Limited	Broad
Termination costs	Inexpensive	Inexpensive	Expensive	Expensive	Inexpensive	Expensive	Very expensive
Time delay	Poor	Poor	Good	Excellent	Good	Very good	Poor
EMI and noise immunity	Poor	Poor	Good	Good	Fair	Good	Excellent
Reliability	Excellent	Excellent	Excellent	Excellent	Excellent	Excellent	Good
System cost	Inexpensive	Inexpensive	Inexpensive	Moderately expensive	Moderately expensive	Moderately expensive	Expensive

(a)

Fiber type	Profile	Diameter cure	Diameter cladding	Diameter coating	Attenuation, dB/km	N.A. (See note)	Bandwidth -3 dB optical	Fiber strength
I	Graded	55 μm (0.0022 in)	125 μm (0.005 in)	1000 μm (0.040 in)	< 8 @ 0.82 μm	0.25	200 MHz	6.9 $\times$ 10 in N/m^2 (100,000 psi)
II	Graded	62 μm (0.0024 in)	125 μm (0.005 in)	137 μm (0.0056 in)	< 7 @ 0.82 μm	0.21	400 MHz	1.7 $\times$ 10 in N/m^2 (25,000 psi)
III	Graded	62 μm (0.0024 in)	125 μm (0.005 in)	1000 μm (0.040 in)	< 7 @ 0.82 μm	0.21	400 MHz	6.9 $\times$ 10 in N/m^2 (100,000 psi)
IV	Graded (modified)	100 μm (0.004 in)	140 μm (0.0056 in)	1000 μm (0.040 in)	< 8 @ 0.82 μm	0.30	20 MHz	6.9 $\times$ 10 in N/m^2 (100,000 psi)
V	Step (PCS)	200 μm (0.008 in)	380 μm (0.015 in)	600 μm (0.024 in)	< 7 @ 0.82 μm	0.22	25 MHz	6.9 $\times$ 10 in N/m^2 (100,000 psi)
VI	Step (PCS)	200 μm (0.008 in)	300 μm (0.012 in)	600 μm (0.024 in)	< 10 @ 0.82 μm	0.26	25 MHz	6.9 $\times$ 10 in N/m^2 (100,000 psi)
VII	Step (PCS)	200 μm (0.008 in)	400 μm (0.016 in)	500 μm (0.020 in)	< 50 @ 0.66 μm	0.22	20 MHz	6.9 $\times$ 10 in N/m^2 (100,000 psi)
VIII	Step (PCS)	300 μm (0.012 in)	440 μm (0.0176 in)	650 μm (0.026 in)	< 7 @ 0.82 μm	0.22	20 MHz	6.9 $\times$ 10 in N/m^2 (100,000 psi)
IX	Step (PCS)	400 μm (0.016 in)	550 μm (0.022 in)	850 μm (0.034 in)	< 7 @ 0.82 μm	0.22	15 MHz	6.9 $\times$ 10 in N/m^2 (100,000 psi)

Note: Numerical aperture (N.A.) values are as defined by Brand-Rex Suppliers. Copies of test methods are available upon request. 100% of all fiber purchased by Brand-Rex has been subjected to the minimum tensile requirements as specified.

(b)

FIG. 21-10 (a) Comparison of fiber-optic versus conventional wire and cable. (b) Fiber-optic selection guide. (*Courtesy of Brand-Rex Co. of Akzona, Inc.*)

well) while other connectors are made specifically to fit Motorola devices. The physical size of a LED may not be the same as its matching detector. Interchangeability of components is yet to be standardized. Another problem area is that of matching cables to emitters and detectors. For example, most companies produce infrared emitter-detector devices, yet most cable companies produce plastic cables for use in the visible range. The designer, therefore, has to carefully select components in order to create efficient low-loss systems. In summary, fiber-optic cables offer the following advantages:

Greater bandwidth
Smaller size and weight
Lower attenuation
Freedom from EMI
Ruggedness and safety
Low cost

In this experiment, cables are compared and different light sources and detectors are used. Components are tested by using practical procedures. In Experiment 22, flux budgeting is considered along with the cutting and polishing of fiber terminations.

TESTS AND MEASUREMENTS

MATERIALS REQUIRED FOR EXPERIMENT

Active Devices	Capacitors
SE4352 IRED	100 pF
SD3443 Detector	47 μF
LED, visible	
7404 Inverter	**Switches**
2N3905 Transistor	DIP, 4PST (2)
TL081 IC amplifier	
IN4148 (3) Diodes	**Fiber Cables**
	50 cm OE1040
Resistors	3 m OE1040
47 Ω	3 m 4001
330 Ω	Fiber cable connectors
510 Ω	
1 kΩ (3)	
8.2 kΩ, 10 kΩ (3)	
100 kΩ	

During this experiment you will evaluate 1-mm fiber cables of different lengths. Also, different emitter sources and detectors will be evaluated with the cables. In the background discussion, general theory was presented. Before the laboratory measurements are taken, some additional information covering practical points will be presented since certain procedures are used in making the laboratory tests.

Measuring Cable Losses

The amount of light power coupled into a fiber depends on:

1. The power output of the emitter (LED/IRED/laser). This is a function of the driving current and radiation pattern of the device.
2. The cable size, its numerical aperture, and the alignment of the fiber core with the emitter.

Measuring the relative optical power coupled into a cable depends on knowing the N.A. and attenuation of the cable coupled to the source and measuring the cable's output in the electrical domain. Small, often

missed factors may affect the measurement. These factors include (1) polished ending of the fibers, (2) distance from the LED to fiber, (3) rotational alignment of cable core with radiant light, (4) type of LED lens, and (5) cleanliness of surfaces.

A power loss of 3 to 6 dB can easily result from the above problem areas. In short runs, power is transferred by means of both the core and the cladding. This makes little difference since all power transferred is quite useful. Distortion of the cable by bending or kinks can also easily attenuate the transmission.

Since plastic fiber cables, such as the SH4001 and OE1040, have higher levels of attenuation, very short lengths, say 50 cm (0.5 m), should be used as a test cable.

Power measurements can be made by comparing the test cable with a second cable which is longer. The power output from the longer cable P_2 is compared with that from the short cable P_1. The attenuation, expressed in terms of decibel loss, is

$$\text{dB} = \frac{10 \log (P_2/P_1)}{L_2 - L_1}$$

When the fiber loss, as measured at the output of the detector's amplifier, is considered, the decibel loss per unit length is

$$\text{dB} = \frac{20 \log (V_2/V_1)}{L_2 - L_1}$$

where L_2 and L_1 are cable lengths and P_2 and P_1 are the respective power outputs for each cable. Typical losses for a 1-mm fiber are in the range of 0.6 to 3 dB/m. When only cables are compared, other losses due to components such as an emitter, detector, or coupler should be held constant.

CONNECTOR AND LED LOSSES

Once the cable loss is known, other losses can be determined. Changes such as in the connectors, LEDs, or the spacing between the LED and the cable are potential loss areas. The absolute power losses are more difficult to determine and require the use of a photometer. The relative output voltage from a detector's amplifier can be compared for different types of LEDs, detectors, and cable-to-LED spacing. Adjustments are made to each component while the shorter cable is used, and the output is maximized. Then assume this to be the standard value. Other outputs are compared to the established values, and the decibel loss is computed. Suppose that the maximum output from the detector's amplifier is 0.3 V. When the LED with a round lens is changed to a flat-surface Fresnel lens, the voltage output drops to 0.21 V. The change in the LED caused a

3-dB loss in the output, hence the cable linkage will exhibit an additional loss. Either the LED power output must be increased, or the gain of the detector's amplifier must be increased in order to compensate for the additional loss.

Further losses are caused by cable fibers that have not been properly terminated. The end of the fiber must be properly cut and polished. If a plastic cable is used, a hot knife is better than cutting pliers. Experimentation is needed to determine how best to finish the end of the fiber. Hand polishing may be better than using a hot knife.

DUTY CYCLE

In testing the fiber, light source, detector, and amplifier, a square wave is best used. Both the leading (high-frequency component) and trailing edges can be observed. The amount of power required will depend on the cable losses, the type of detector, and the gain of the receiver. The power delivered in a square wave pulse train depends on the duty cycle. The duty cycle is determined by the ratio of the pulse period (t_1 — the width of the pulse) to the time T between the leading edges of two consecutive pulses. Figure 21-11 shows the time period for determining the duty cycle and peak power.

Since t_1 is a percentage of the period T, the duty cycle is often expressed as a percentage. A 20 percent duty cycle indicates that peak power pulses are present for 20 percent of the time. A 50 percent duty cycle represents equally divided pulse periods, and the average power is one-half the peak power. The *average power* P_a of a pulse cycle equals the peak power P_p multiplied by the duty cycle. If the peak power is 20 mW and the duty cycle is 20 percent, the average power is

$$P_a = P_p \times \text{duty cycle}$$
$$= 0.020 \text{ W} \times 0.2 = 4 \text{ mW}$$

At a 50 percent duty cycle, the average power is 10 mW. If a light power meter (photometer) is available, the power can be measured at the different locations in the system.

An important element in power transfer involves using components designed to operate on the same wavelength. An infrared emitter feeding a cable which is designed for operation at a visible wavelength will cause an attenuation in the power transfer. In the laboratory evaluation, different cable lengths will be used with light emitters of different wavelengths. Losses will be observed.

FIBER CABLE COMPARISON AND LOSSES

In the following tests, two 3-m cables are compared with a 50-cm cable, and the losses are computed. In the test, a Crofon cable (by Dupont) is evaluated. The published losses are stated as follows: model 1040, approximately 2 dB/m; OE1040, 0.9 dB/m. The ESKA 4001 cable has a loss of 0.6 dB/m.

In the first test, a 3-m length of OE1040 cable is compared with a 50-cm length of similar cable. Note that the emitter and detector used in the tests are more sensitive in the near-infrared range while the cable has its minimum losses in the visible range.

An unmodulated light source is transmitted by an IRED and received by a photodiode and then amplified.

1. Connect a 50-cm length of Crofon OE1040 cable, with plugs on each end, between the IRED and the photodiode. Figure 21-12b and c shows the circuits to be used. In the amplifier use a 10-kΩ resistor for R_2 and a feedback resistance of 1 kΩ. The amplifier provides a gain of approximately 2. Use a DMM to read the output voltage at TP_5.

NOTE: The receiver circuit shows the use of switches for changing the feedback resistor values. Close 3d and 4d. For the photodiode load, close 1d and open 2d.

On the transmitter close 1a and 2a, then open 3a and 4a. What is the voltage at TP_3?
2. Change the 50-cm cable. Insert a 3-m length of ESKA 4001. Again, what is the voltage at TP_3?
3. Change the 4001 cable to an OE1040. What is the voltage at TP_3?
4. Compute the cable loss per meter for type 4001 and OE1040 cable. What is the loss per meter for each?
5. Use a photometer to measure the actual power being transmitted by each cable — P_o for 50-cm cable, 3 m of 4001, and 3 m of OE1040.
6. Which cable (4001 or OE1040) exhibits the lowest loss in the infrared region?
7. Disconnect the infrared emitter and replace it with a visible LED (Fig. 21-12d). Since the LEDs are physically different (one with a lens and one with different operating voltages, etc.), only relative differences between cables are considered. With a feedback resistance of 19.2 kΩ (3d and 4d are open), measure the output V_o at TP_5 with a 50-cm cable. Check the voltages of both 3-m cables.

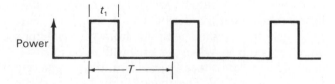

FIG. 21-11   Determining the duty cycle.

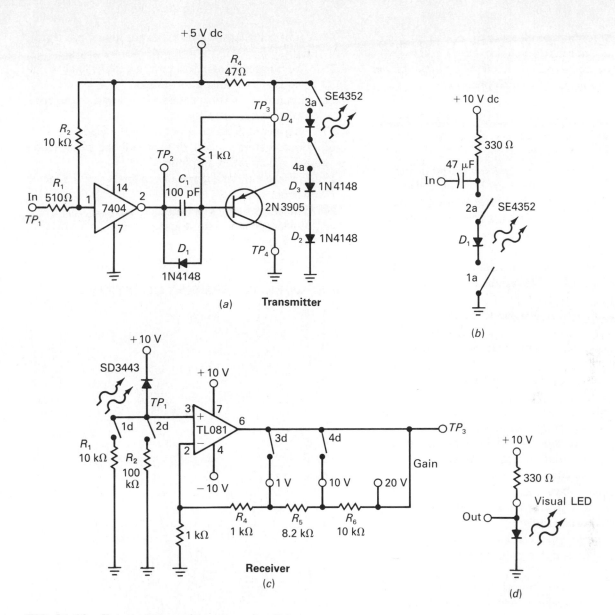

FIG. 21-12 Transmitter and receiver circuits.

Which 3-m cable shows the least loss in the visible region? Use an input resistance R_2 of 100 kΩ.

Bandwidth Measurement

In the following procedure, a square wave will be transmitted through a 50-cm cable, and the effects are observed when the load resistor in the photodiode circuit is varied. The experiment will confirm that when sensitivity (responsivity) is increased, the bandwidth is decreased.

8. In the receiver use a 100-kΩ resistor R_2 for the load. Connect a function generator, square wave output, to the input (In) of the circuit shown in Fig. 21-12a. Increase the signal voltage until a firm

pulse train is observed at TP_3 of the transmitter. There should be an output from TP_5 of the receiver. Reduce the receiver gain if it oscillates or if it overloads. Observe the square wave as the frequency is increased from 1 kHz. The leading edge will start to roll off with a high-frequency loss. In what frequency range does this occur? Change the photodiode load to 10 kΩ (R_1). Where does the rolloff occur?

9. Is the rolloff caused by the cable? Change to the 3-m 4001, and determine whether it affects the rolloff. If the photodiode load resistor were decreased to 1 kΩ, what would you expect to happen to the bandwidth?

10. How much loss, in decibels, would you expect if type OE1040 cable were used for 1 km?

11. Use a photometer to measure the power output P_o from the 3-m OE1040 cable when a 1-kHz signal has a duty cycle of 50 percent. What is the power output when the duty is 25 percent?

REVIEW QUESTIONS

Test your comprehension of the subject by answering the following questions. Answer True or False on a separate piece of paper.

1. Considering overall cost, the fiber cable and connectors are more costly for a 5-mi length than copper.
2. From technical data provided, the frequency range of ESKA cable exhibits the least loss at 800 nW.
3. The cleaving of a fiber-optic cable refers to its length.
4. Light travels through a fiber cable faster than it does in air.
5. Fiber-optic cables are rated in terms of decibels per kilometer.
6. When a fiber cable is made very small in diameter, you can expect a greater loss.
7. As the N.A. of a cable is made smaller, the light wave travels straight through the cable.
8. In general, glass fiber cables have a lower loss than plastic fiber cables.
9. Repeater stations are closer together in fiber cable links than in wire links.
10. A 1-mm cable conducts one light mode at a time.
11. Fiber cables are not immune to high ambient temperatures.
12. Fiber cables have a frequency response.
13. The light generally conducted through a glass cable usually operates in the infrared region.
14. Light travels through the larger fibers in a zigzag path.
15. The frequency response of the cable is likely to be wider as the cable gets smaller.
16. Numerical aperture refers to the angle of the cone at the entrance to the fiber.
17. There is no light conduction in the cladding.
18. Photometers consist of a broadband photodiode and a calibrated amplifier.
19. Photometers are generally calibrated at a specific frequency.
20. The smaller the diameter of the fiber cable, the lower will be the light loss.

ANSWERS TO REVIEW QUESTIONS

1. F
2. F
3. F
4. F
5. T
6. F
7. T
8. T
9. F
10. F
11. F
12. T
13. T
14. T
15. T
16. T
17. F
18. T
19. T
20. T

FLUX BUDGET DESIGN AND OPTOCOMPONENTS

SCOPE OF STUDY

The high-energy output of light-emitting transmitters and the use of highly sensitive receivers are offset by the losses that exist between the components making up a transmission system. This experiment covers the methods used in computing flux losses and gains in a communications link and the way to properly attach connectors to fiber-optic cables, in order to reduce link losses.

OBJECTIVES

Upon completion of the experiment and the evaluation of the laboratory procedures, you will be able to:

1. Describe the importance of determining the flux budget of an optical link and the effects of responsivity
2. Specify which components in the link contribute the most to the loss
3. Describe and demonstrate how to prepare and arrange optical components in order to minimize system loss

BACKGROUND

The linkage between the radiant light from the emitter of a transmitter and the photodetector of a receiver contains a variety of losses. If the emitter were placed next to the detector, there would still be losses. Losses caused by reflection, scattering of light through different materials, and frequency-response differences are but a few of the contributing factors. If the emitter is transmitting its light through a fiber-optic cable, even greater losses will exist. The flux budget, or loss budget as it is sometimes called, is a method of comparing and computing the ratio of the transmitter's radiant light flux to the receiver's input sensitivity, as compared to the total loss in the optical link. For each component, such as fiber cables, the alignment of the emitter and detector, frequency responses of devices, physical spacing between components, and numerical apertures are loss-contributing factors. The following is a listing of where losses often occur and the decibel loss contributed.

Areas of Loss	Loss Contribution (dB)
Emitter-to-fiber N.A.	1–8
Emitter-to-fiber space area	0–2
Transmitter gap loss	0.5–4
Transmitter misalignment	0–4
Entry Fresnel loss	0–0.5
Fiber attenuation (depends on length)	0–over 100
Fiber splices	1–6
Exits Fresnel loss	0–0.5
Receiver gap loss	0.5–4
Receiver misalignment	0–4
Detector Fresnel loss	0–0.5
Fiber-to-detector N.A.	0–2
Fiber-to-detector area loss	0–0.5

The possible total loss in the optical path is 3 to over 100 dB. The optical loss in the system is the sum of all losses in the transmission linkage. The difference between the transmitter-to-receiver flux ratio (FR) and the system's insertion losses (IL) is the *flux margin*. The flux margin M must always be greater than zero (otherwise power losses exceed power transfer). The identification and the quantifying of each loss enable the optoelectronics designer to calculate the required transmitter power and/or the receiver sensitivity necessary to compensate for the losses.

The LED or IRED output of the transmitter is usually rated in terms of microwatts, and the optical flux can be expressed in terms of decibels or dBm. The following symbols are used in flux calculations:

Transmitter output flux	$\phi_o T$
Receiver input sensitivity	$\phi_o R$
1000-μW reference (1 mW)	ϕ_o
System flux ratio	$\propto$FR
Flux margin	$\propto M$
System insertion loss	$\propto$IL

For ease in making calculations all power ratings are referenced to 1 mW (1000 μW). The dBm is used to express the ratio of the radiated flux to the standard (1000-μW) flux ϕ_o. The transmitted output flux is

$$T \text{ (dBm)} = 10 \log \frac{T \text{ (mW)}}{\phi_o}$$

An IRED has an output of 2 mW. The power, expressed in terms of dBm, is

$$10 \log \frac{2000}{1000} = 3 \text{ dBm}$$

A LED has a power output rating of 150 μW. Its dBm value is less than 1. Thus

$$dBm = 10 \log \frac{150}{1000} = 10 (-0.823)$$
$$= -8.23 \text{ dBm}$$

If both the transmitter's radiated power and the receiver input sensitivity are expressed in dBm, the *system flux ratio* (SFR) is ϕT dBm $- \phi R$ dBm. All other losses will reduce this ratio. The flux margin is the difference between SFR and other insertion losses. The flux margin must always be greater than zero. From a practical point of view, to measure the flux losses, the radiated power output of the IRED should be measured first. The output power from a cable of length L includes the losses at the connectors and in the cable. The difference between the IRED output power and that which is available at the detector is due to the flux margin. Estimates can also be made by knowing the cable loss and by assuming typical losses at the connections of the emitter and detector.

RESPONSIVITY

The term *responsivity R* refers to the ability of a photodiode to increase its back-biased current as a result of an increase in radiant flux. The responsivity of the photodiode is rated in microamperes per microwatt within a certain wavelength range. The Motorola photodiode MFOD100 has a typical rating of 0.5 μA/μW. When the radiated light at the cell, for example, is 2 μW, the diode will produce a current flow of 1 μA (0.5 $\times$ 2 μW). The peak response of the MFOD100 is approximately 820 nm. Figure 22-1 shows the diode connected to a load resistor in which a signal voltage is developed. A current flow of 1 μA through a resistor of 10 kΩ will produce a signal voltage of 10 mV. The larger the load resistor, the higher the signal voltage developed but the lower the bandpass. The value of R is determined by establishing a signal voltage of 10 mV.

Although the responsivity of the photodiode is in the range of 0.2 to 0.6, the phototransistor has a rating in the range of 100 and the photo-Darlington's rating is nearer to 400 than 600.

The photodiode's capacitance is small, and its response time may be 50 ns or less. The phototransistor's rise time is longer and is in the range of 1 to 20 μs. It is far slower than the photodiode. The Darlington is still slower, with a rise time of 10 to 100 μs. High-speed switching, therefore, requires the use of photodiodes. The photodiode's lower gain has to be made up by a high-gain, low-noise, wide-bandwidth amplifier. In view of the losses in the fiber line and connectors, much depends on the responsivity of the receiver's detector and the power output of the transmitter. Properly attaching connectors to a fiber cable can reduce the flux loss and increase the flux margin. The following information provides some highlights on the handling of 1-mm fiber-optic cable and the use of typical connectors.

ATTACHING CONNECTORS

The main objective in preparing the ends of a fiber cable is to reduce light reflections and scattering. Both plastic and glass fibers should be handled according to the manufacturer's recommendations. The following procedures were supplied by Nissho-Ivai American Corp. for the handling of their (ESKA) plastic fibers (1 mm in diameter). The attaching of Optimate DNP plastic connectors was supplied by AMP Co. ESKA is a registered trademark of the Mitsubishi Rayon Co., Inc., Japan.

The Dupont cable OE1040 is 1 mm in diameter, and it is similar to the ESKA in size and can be fabricated in a similar manner. The OE1040 has a decibel loss equivalent to that of the ESKA cable (4001).

GENERAL PLASTIC FIBER TERMINATION PROCEDURES (BY ESKA)

The amount of time spent in finishing plastic fiber ends depends on what the user is attempting to accomplish with the plastic fiber link, e.g., transmission length.

The termination methods may range from polishing to contact-heat lensing, to hot-blade cutting. User experimentation is recommended in choosing the most suitable method which will meet the economic and operational parameters set up for the link.

To reduce the amount of light loss, e.g., scattering and angular losses, at the ends of plastic fibers, the surface of the tip should be vertical to its horizontal axis and ideally should be of mirrorlike smoothness.

The end of the fiber should also be free of dust

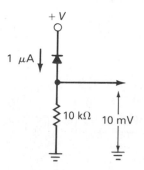

FIG. 22-1 Current flow responsivity.

particles and stain which cause absorption losses. For cleaning, ethyl or isopropyl alcohol on a soft swab can be used sparingly to clean the fiber's tip.

A small 10X magnifying glass can be used to inspect the finished surface of the fiber.

1. VERTICAL CUTTING OF THE PLASTIC CABLE
   a. If the polishing or contact-heat lensing termination methods are used, it is suggested that the plastic cable end (prior to stripping off of the cable jacket) be cut vertical to the fiber's horizontal axis.
   b. This vertical slice will eliminate deep or rough ridges on the face of the fiber tip and assist in attaining a mirrorlike finish.
   c. Always keep the blade clean and sharp. Do not use blades which are coated with Teflon or other substances. Such coatings may cause a light loss at the fiber face.

2. POLISHING THE FIBER ENDING
   a. After completing the vertical cutting, follow the connector manufacturer's instructions in preparing the fiber for insertion into the ferrule or plug. Do not scratch or damage the fiber's cladding since light loss may occur.
   b. Do not use epoxy adhesive which contains amines. Epoxy adhesives which are thiol-based are acceptable, but only if they are really necessary.
   c. Use a polishing fixture to hold the fiber, both stationary and vertical to the polishing surface.

3. POLISHING STEPS FOR PLASTIC CABLES (SUGGESTED BY ESKA)
   a. First, polish with a no. 600 grit sandpaper. Wet the sandpaper with a little water. Use a figure-eight polishing motion during all polishing.
   b. Follow the no. 600 grit by polishing with a 12-μm lapping tape.
   c. Finish the fiber surface with a 0.5-μm lapping tape.
   d. Clean ends with alcohol.

4. CONTACT-HEAT LENSING METHOD (BY ESKA)
   a. Acrylic (polymethyl methacrylate) is amorphous in molecular structure (i.e., it has a noncrystalline makeup), and so no melting point is defined for it. Instead, the term *flow point* or *contact-forming point* is used in relation to plastic fibers.
   b. The heat-contact-forming temperature is 100 to 140°F (37.8 to 60°C). By contact-forming range, it is meant that when a plastic fiber is pressed against a heated surface, the fiber's tip will soften, "flow" away from the heat source, and tend to form a lens. Lower temperature levels require greater pressure against the heated surface and a longer contact time. The longer the contact time, the larger the lens will become.

Note that the aperture in the plug's tip must accommodate the enlarged fiber tip or lens. That is, if the lens cannot recede into the aperture, it may interfere with the positive locking action of the plug.

Preliminary studies indicate that the AMP DNP connector plug with its V-grooved split ends is suitable for the plastic contact-heat lensing method. By experimentation, the link designer can determine the amount of plastic fiber tip that can be lensed into the plug's aperture. If the heated surface is flat and smooth and has a mirrorlike finish, the fiber's face will also exhibit a similar finish. A stainless-steel plate is recommended.
   c. Contact-heat lensing procedures
      (1) Make a vertical cut as indicated.
      (2) Follow the connector manufacturer's instructions for stripping and inserting the fiber into the plug; however, allow a little excess fiber for the lensing.
      (3) The temperature range of the heated surface should be 100 to 140°F, and the contact time (fiber to heat source) should be between 0.2 and 0.4 s.
      (4) With practice, the excess amount of fiber to allow for lensing, the ideal temperature, and the contact time should be more easily determined.
      (5) Take care not to melt the plug tip onto the fiber face. For example, the DNP plug (AMP Co.) has a melting point higher than the flow point of plastic. But caution should still be exercised since both are plastic and both materials may melt.

5. HOT-BLADE CUTTING METHOD
   Preliminary results using this method show that the ideal temperature of the cutting blade is in the range of 160 to 180°C. Higher blade temperatures will produce deep and/or sharp ridges which will cause undesirable results on the fiber tip face. Such ridges will cause a light loss. A 25-W soldering iron with a hot knife may produce a tip temperature as high as 400°C. Again, care must be exercised.

   Use a blade with a beveled edge. Keep the blade clean. If the temperature of the hot knife can be controlled, hot-blade cutting will produce good results on ESKA cables. Avoid melting the plug or ferrule tip onto the fiber face during hot-knife cutting.

6. USING THE DNP OPTIMATE CONNECTORS
   The AMP Optimate DNP fiber-optic simplex and duplex plugs, splicer-terminated for 1000-mm plastic fiber (single- and dual-channel cables), are shown

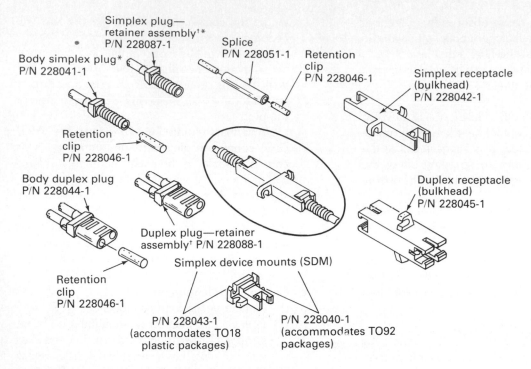

FIG. 22-2 AMP plugs.

*Can mate with both simplex and duplex receptacles.
†Includes retention clip.

in Fig. 22-2. Two items are recommended for use in trimming the fiber after plug termination, namely, a hot knife and a cutting tool fixture.

7. PLUG TERMINATION (SIMPLEX AND DUPLEX) Omit step *a* of this procedure when using P/N228087-1 or P/N228088-1.

 a. Insert the retention clip into the plug body. The V-notched end of the clip goes first, as shown in detail A of Fig. 22-3. The opposite end of the clip is flush with the plug end.

 b. Strip approximately ⅛ to ¼ in of cable jacket, using no. 20 AWG gage on wire strippers.

CAUTION: Do not cut or nick fiber.

NOTE: When using wire strippers, be sure to set the strippers to no. 20 AWG wire size before stripping cable. When using a wire stripper for the first time, you can use a section of fiber as a test gage.

 c. With pliers (for ease of insertion) push stripped cable into the plug until the cable bottoms and excess fiber is exposed.

NOTE: Cable jacket should not enter the slotted portion of the plug. See detail B of Fig. 22-3.

 d. Pull back slightly on the cable to establish cable retention. The fiber is now ready for trimming.

 e. Slide the terminated plug into the cutting tool

fixture, as shown in Fig. 22-4. Ensure that the face of the plug is flush with the face of the fixture.

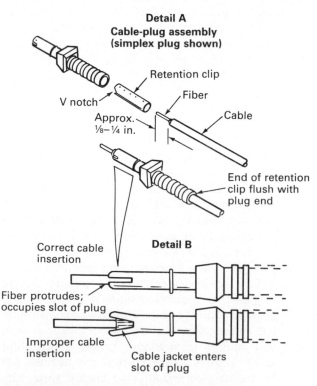

FIG. 22-3 Optimate cable plug assembly.

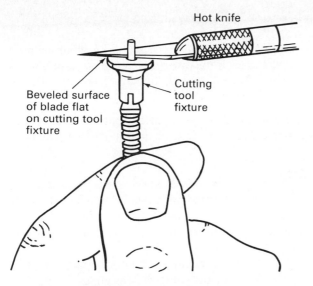

FIG. 22-4 Hot-knife cutting of fiber.

f. Hold the plug/fixture assembly at the base of the plug as shown.

=== **WARNING** ===

KEEP FINGERS AWAY FROM CUTTING TOOL FIX-TURE WHEN TRIMMING FIBER WITH HOT KNIFE. ALLOW FIXTURE TO COOL BEFORE REMOVING THE PLUG.

g. Angle the hot knife (recommended: Weller SP23HK) so that the blade bevel is flat on the cutting tool fixture. With light blade pressure on the fiber, cut off the excess fiber, making it flush with the face of the plug assembly. See Fig. 22-4.

NOTE: The proper heat zone of the hot knife blade should be determined by usage. The heat zone becomes colder toward the blade tip. If fiber melts, the zone of the blade directed on the fiber is too hot. If fiber snaps off, the zone of the blade is too cold.

h. Allow the fixture to cool from blade heat transfer (if any); then remove the plug from the fixture. Now the plug assembly can be mated with any of the receptacles.

8. OPTICAL FIBER SPLICE
   To connect two optional fibers using the AMP Optimate DNP fiber-optic splice, perform the following steps for each optical fiber.
   a. Slide the retention clip on the cable with the V-notched end of the clip approximately ⅛ to ¼ in away from the end of the cable (Fig. 22-5).
   b. Using the hot knife, cut off the cable and make the fiber flush with the V notch end of the clip (Fig. 22-5). For ease of insertion into the splicer, grip the cable directly behind the retention clip with pliers and push the retention clip-cable assembly into splicer until the edge of the retention

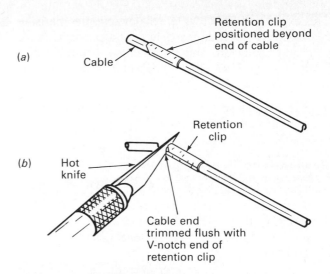

FIG. 22-5 Attaching clip.

clip is flush with the edge of the splice. See Fig. 22-6.

TESTS AND MEASUREMENTS

MATERIALS REQUIRED FOR EXPERIMENT

Active Devices	Resistors
SE3452 IRED	820 Ω
SD3443 Photodiode	1 kΩ (2)
TL081 IC amplifier	8.2 kΩ
	10 kΩ (2)
	100 kΩ

Miscellaneous
50 cm of 1-mm fiber cable with connectors
   228087-1 (2)
AMP two retention clips, 1 mm cable, plugs (2)

Accessory Tools
Hot knife, polishing kit (SIC376)
Stripper
Soldering tool
Photometer (SIC374)

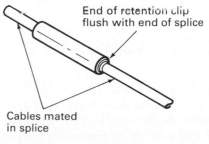

FIG. 22-6 Coupling.

This experiment provides two major activities, namely, the evaluation of the responsivity of a photodiode for gain and bandwidth and the practical assembly of a connector to a fiber cable. Flux budget measurements require the use of a photometer.

1. The circuit shown in Fig. 21-12*a* and *b* is also to be used in this experiment.

Responsivity Measurements

2. The receiver feedback resistor should be connected to provide an amplifier gain of 10. The light emitter should be the IRED. The detector in the receiver is a photodiode with a detector load resistor value of 100 kΩ (to start).

   For setting the emitter current, use a resistor value of 330 Ω. With a +10-V supply the emitter current should be 25 to 30 mA. Connect a 50-cm fiber cable between the emitter and photosensor.

   Feed the 47-mF capacitor to the emitter with a 1-kHz sine wave set to approximately 0.2 V. Use a dual-trace scope to monitor the function generator and the receiver output at TP_5. Adjust the input signal level until the receiver's output is equal to 1 V p-p. Record V_{in}.

3. What is the receiver voltage output V_o when the photodiode's load resistor is changed from 100 kΩ to 10 kΩ?

Determine the Light Level on the Detector

4. Start with a 1-V output (100-kΩ input) at the receiver and with a receiver gain of 10, determine the following data:
   a. Input voltage V_{in} (computed for stage gain) to the receiver amplifier
   b. Current I_R through the 100-kΩ resistor
   c. The responsivity ratio value for an SE3452 of 0.4 per milliamperes per microwatt.
   d. The microwatt level
   e. The above is an indirect way of determining the light power which is radiated on the photodiode. The direct method requires the use of a photometer. If a photometer is available, connect the fiber cable between the emitter and the photometer, and record the power output P_o from the cable.

Bandwidth versus Photodiode Load Measurement

5. Maintain the generator's output at 1 V p-p. Start with a detector load resistor of 100 kΩ. At 1 kHz, the receiver output should be 1 V. Determine the upper bandpass (3-dB rolloff) by increasing the frequency of the generator until the receiver's output

falls to 0.707 V. In what frequency range does this occur?

6. Check the high-frequency response when using a 10-kΩ load. If necessary, increase the generator's signal. The endpoints, however, are still 0.707 times the output voltage.

NOTE: The amplifier's response is not flat. In fact, you might find a gain increase at the higher frequencies.

Cabling

7. Construct a 50-cm fiber cable using a 1-mm plastic cable and two terminal plugs. Three different methods of terminating a plastic cable are evaluated.
   a. Cut both ends with a knife or cutting pliers, and do no other polishing of fiber ends.
   b. Trim the cable ends with a knife, and smooth the ends by lensing against the side of a soldering iron (not at or near the tip—use 100 to 140°F (37.8 to 60°C).
   c. Cut the cable with a hot knife following procedures described in the background discussion. Be sure to clean the ends with alcohol after polishing them. Check your results by using either the amplifier and photodiode (Step 2) or the preferred method of using a photometer. Record the reading for all three methods used.
   d. Try polishing the ends of the cable with polishing rouge. The Science Instruments Co. provides a polishing kit, model S378, a fiber optic inspection tool, S376, and a photometer, S374. The polishing method is easier to use than the hot knife method.

8. Which method of preparation produced the least loss?

9. Use a 50-cm fiber cable between the emitter and photometer. Measure the value of P_o.

10. Change to a 3-m cable (1040) and again measure the value of P_o.

11. While reading the output power, rotate the plug at the emitter to determine whether there is any difference in how the light enters the cable.

12. What happens if the fiber cable ending has a gap between the cable and emitter? Move the plug slowly out from the emitter socket and observe the power reading. From the previous observations, it was found that losses occur in the connector, as a result of air gaps, in the method of terminating a cable, and according to the type of cable.

REVIEW QUESTIONS

On a separate piece of paper, complete the following statements with the appropriate word or words.

1. Flux margin is the difference between _____ and _____ .
2. The symbol "dBm" is the ratio of _____ as referenced to _____ .
3. Responsivity is a numerical value for the ratio of _____ to _____ .
4. The responsivity factor must be greater than 1, True or False? _____ .
5. By flux budgeting the engineer can determine the required _____ output and the needed gain of the receiver.
6. The load resistor of the photodiode controls the _____ and _____ of the receiver.
7. A lens on the emitter _____ the decibel gain of the links.
8. A highly polished terminal ending will _____ flux losses.
9. A Crofton type OE1040 cable could have a loss of _____ dB/km.
10. A photometer is usually calibrated in terms of _____ or _____ .

ANSWERS TO REVIEW QUESTIONS

1. The ratio of the transmitter/receiver transfer and the sum of the other losses
2. Power, 1000 μW
3. Microamperes to microwatts
4. False — the transistor detector could be 50 to 100
5. Emitter
6. Gain, bandwidth
7. Increases
8. Reduce
9. 2000
10. Milliwatts or microwatts

PART SIX

APPLICATIONS TO ANALOG TELECOMMUNICATIONS

SCOPE OF STUDY

This part discusses a variety of methods of transmitting and receiving signals. Different types of modulators and demodulator circuits are used to transmit information over a fiber-optic cable.

OBJECTIVES

Upon completion of the experiments in Part 6 and evaluation of the data, you will be able to:

1. Describe how analog signals can be transmitted via fiber-optic cables
2. Suggest applications for the principles involved in the circuits
3. Test and evaluate modulators and demodulators
4. Communicate your understanding of the different modes of modulation with other technical and nontechnical people
5. Research optoelectronic components for use in telecommunications
6. Maintain and repair light communication equipment using optodevices, given additional information and training for the overall operation of the equipment
7. Read and comprehend technical articles which describe the use of fiber-optic links

The circuits included in Part 6 are designed to transmit or control analog signals, i.e., linear varying signals, using optical devices. The circuits presented are typical of those used in wire or wireless communications, but they are adapted to optoelectronics techniques.

Amplifiers and the modulation of carrier signals by AM, FM, pulse-code modulation (PCM), and other methods are included in the study. The transmission of analog signals is encountered in process instrumentation, industrial controls, telephony, and voice and video transmission. Problems such as attenuation, noise, distortion, and fidelity, which are observed in other systems, are also present in optical systems.

Light beam communications have been in use for some years. The fidelity of early light communication systems was poor since the modulation of tungsten lamps was the most popular technique. The frequency response of tungsten lamps is poor, hence fidelity suffered. The LED, however, can be modulated at much higher frequencies, and the quality of reproduction is greatly enhanced. Although the fidelity is improved, the light output of the LED or IRED is far less than the tungsten lamp. This reduction in illumination requires that the sensor (detector) circuitry provide additional gain. The Darlington-type detector or PIN diode with an IC amplifier is popularly used in this application. To further increase the light output of the LED, it is usually designed into a pulse AM or FM type of circuit. Power dissipation in the LED limits its ultimate distance of transmission

103

and frequency range. Where transmission distances exceed 10 to 20 ft or where the transmission is to be immune from electromagnetic interference (EMI), fiber-optic cables can be utilized. Such cables are also used in short links of 1 to 3 ft.

Typical frequency responses exceed 100 kHz. High frequencies (above 100 kHz) can be transmitted by using high-frequency and high-current pulse generators, quality IRED radiators and sensors, and low-loss fiber cables. At both the lower and higher frequencies, frequency and pulse modulation can be used.

Although reference has been made to data transmission via fiber links, the requirement may only call for the separation or isolation of the signal source from the remainder of the equipment. In medical applications the isolation of a patient from earth ground or from equipment electrical potentials prevents electric shock hazards. Couplers are effectively used in such applications.

The emitting diode (the transmitter or radiator) can be modulated in two basic ways. The direct current of the diode can be ac- or pulse-modulated. This means that the dc level is changing at an audio rate (similar to AM), or the LED can be driven by an alternating or pulsed current whose frequency is pulse-width-modulated or pulse-coded. The two systems are ac modulation of a direct current or ac/pulse modulation of an alternating current.

In the circuit applications that follow, a variety of assignments, using different forms of modulation, are presented.

MODULATING A LED OR IRED

SCOPE OF STUDY

The general characteristics of LEDs are discussed, and practical circuits are evaluated in the laboratory. Also covered is the description of the laser diode and the physical characteristics of different types of light-emitting diodes.

OBJECTIVES

Upon completion of the assignments and evaluation of the data, you will be able to:

1. Describe several methods of modulating a LED or IRED
2. Suggest applications for the principles involved in the circuits
3. Test and evaluate LED modulator circuits
4. Communicate your understanding of the function of light modulators to other technical and nontechnical people
5. Research LED components for their characteristics
6. Maintain and repair industrial/commercial equipment using LEDs, given additional information and training on the overall operation of the equipment
7. Read and comprehend technical articles which describe the use of light-emitting devices

BACKGROUND

The LED is well known to almost everyone. The applications of the LED in calculators, games, clocks, and dashboard panels makes it important. In many applications the LED has replaced the tungsten lamp. The LED used as a display light is bright, colorful, and compact; has low power consumption; and is economical.

Some LEDs are designed for use in data and analog transmission systems. These diodes have four main characteristics:

1. Output power
2. Wavelength
3. Speed
4. Emission pattern

The higher the emitted output power that can be effectively coupled, the lower will be the receiver's power gain circuitry. The optical lenses are reduced, and greater losses can be accommodated elsewhere in the system.

The simplest of the LEDs are the epitaxially grown surface and the planar diffused surface devices. Sketches of their physical design are shown in Fig. 23-1.

The epitaxially grown LED is made of silicon-doped gallium arsenide. This device is multilayered with P- and N-type materials. The wavelength of the light is in the 940-nm range, and typically it radiates approximately 3 mW of power when the forward current I_f is 100 mA. The turn on-off speed is in the order of 150 ns, which is relatively slow. The scattering of emission, which is nondirectional, makes it a poor choice as a light source for fiber-optic cables. Fiber cables require a narrow light beam, since the cable diameter is small, and the wavelength of transmission is in the range of 700 to 750 nm. In this frequency range, a specific fiber cable has an attenuation of 10 dB/km. Figure 23-2 shows the attenuation curve for a typical fiber cable. As can be seen, a LED with an emission of 875 nm would be desirable since the fiber attenuation is 7 dB/km in this part of the spectrum and losses are minimized. It should be kept in mind that cable losses are being low-

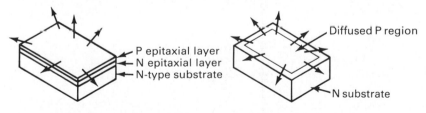

Epitaxially grown surface LED   Planar-diffused surface LED

FIG. 23-1   Epitaxially and planar LEDs.

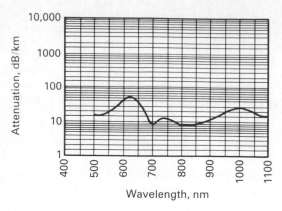

FIG. 23-2   Attenuation of a fiber cable.

ered quite rapidly. Some 9-μm glass fibers have a loss of 0.1 dB/Km.

As noted above, the turn on-off time is slow — 150 ns. In pulse transmission, typically data pulses could be at speeds of 20 to 25 ns. A system with a speed of 100 to 150 ns could not respond to such fast speeds (20 to 25 ns).

The nondirectional feature of the epitaxial chip would require that the LED produce a higher power output since much of the irradiant light would not impinge on the fiber cable core. The emission pattern is referred to as being *Lambertian*, meaning that the radiation is analogous to reflected light radiating off a sheet of paper.

A second type of chip production is the *planar diffused surface LED*. Typical power outputs are 500 μW at a wavelength of 900 nm. The wavelength is adequate, and the speed is 15 to 20 ns. The emission, however, is still Lambertian.

Two types of LED designs that more effectively couple into fiber cables are the planar heterojunction and the edge-emitting diode. An effective LED currently being made is the tertiary crystal of aluminum gallium arsenide. It is used extensively since it radiates in the 820-nm range. In the heterojunction device, the current is concentrated in a very small area. This increases current density and makes for a brilliant spot which focuses on the fiber cable.

When LEDs are to be used in telecommunication circuits operating in the range of 50 to 200 MHz, planar LEDs are impractical because of their slow speed and wide light spread. Cables designed for these frequencies are generally 50 to 65 μm in diameter. To make up for system losses, the LEDs have to operate at higher currents (and temperatures). A newer development in LED design is the etched-well or Burris diode (Burris and Dawson of Bell Laboratories).

The radiation from the well-like structure flows through a slit of approximately 0.025 mm. The emitting area is very small; therefore, the production cost is higher. The edge-emitting diode is highly directional,

with light being radiated from a slit rather than a circular area. The following is a list of key LED features:

1. Forward operating current I_F
2. Operating voltage
3. Reverse peak breakdown voltage
4. Spectrum
5. Output power
6. Turn on-off time
7. Reverse leakage current
8. Cone of radiation

AC MODULATION

The dc forward current (I_F) flow in a LED, when used in a light modulator, serves as the carrier of information. This current can be increased or decreased in amplitude by a linear or pulsed signal. The operation is analogous to amplitude modulation in wireless transmission. The dc carrier in light communications is amplitude-modulated at an audio or pulsed rate. Figure 23-3 shows a basic circuit for capacity coupling an ac signal to the dc circuit which provides the light.

The resistor R and the dc supply establish the forward current flow through the LED. Typically, this current ranges from 10 to 50 mA. The LEDs' operating voltage is approximately 1.2 to 1.6 V. The ac signal voltage is capacity-coupled to the diode. The size of the capacitor depends on the ac frequency and current flow of the diode. The operating resistance of the diode is approximately

$$R_{dc} = \frac{V_o}{I_f} = \frac{1.6}{0.03} = 53\ \Omega$$

(Assume $I_f = 0.03$ A.) Assume a current change of 2 mA, with modulation. This would represent a change in current of 6.6 percent.

The size of the capacitor is determined from

$$T = RC \qquad F = \frac{1}{T} = \frac{1}{RC} \qquad C = \frac{1}{FR}$$

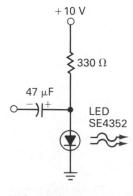

FIG. 23-3   AC coupling to a LED which has a direct current flow.

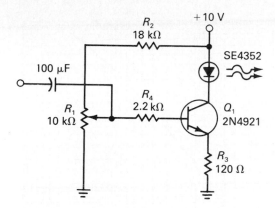

FIG. 23-4   LED driver amplifier.

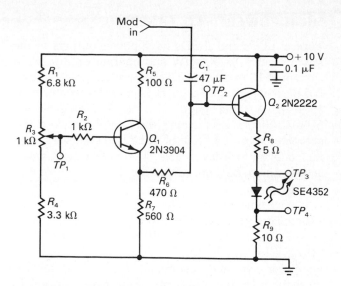

FIG. 23-6   LED modulator.

With operation at 40 Hz, the minimum value of C is 470 μF. The required ac peak voltage needed to cause a 2-mA change (when operation is at 30 mA dc) is determined empirically by:

1. Viewing the LED ac voltage
2. Increasing the ac drive until distortion just starts and then operating below this point
3. Taking half of the peak-to-peak voltage and adding it to the dc voltage of the diode
4. Determining the LED current by using the series resistor voltage drop (the amount of increased current is a percentage of the initial current).

Figure 23-4 shows how a transistor driver can be added so as to improve the modulator. The LED serves as the transistor load. Resistors R_1 and R_2 determine the transistor bias and collector current I_C. Since $I_C = I_F$, a change in R_1 can be used for setting the LED current. Resistor R_3 is a LED current limiter, and Q_1 is an amplifying transistor. Resistor R_4 is the base load resistor.

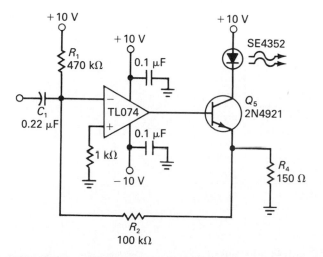

FIG. 23-5   LED driver and preamplifier.

Figure 23-5 shows the use of an IC as a preamplifier. For the component values shown, the voltage at the emitter of Q_1 is $+4.5$ V and the base is $+5$ V. Here $V_{BE} = 0.5$ V, and I_C (I_F) equals approximately 30 mA. The IC is one-quarter of a TL074. An additional section of this IC can be used as an audio amplifier to provide additional gain for a microphone.

The circuit shown in Fig. 23-6 provides a variable direct current and a method of modulation. The circuit has numerous applications in carrier modulation and pulse transmission. The transistor Q_2 sets the current level of the LED. Transistor Q_1 is an amplifier, and varying the base bias causes the collector to control the base of Q_2. The more the center of the potentiometer approaches ground, the higher will be the LED current. At $+0.9$-V base voltage (Q_1), the LED current is 40 mA. At $+1.30$ V the LED current equals zero.

The ac modulating signal is fed to either Q_1 or Q_2 through a coupling capacitor. The circuit will accept a sine wave, square wave, or pulses. For the values shown, the maximum current setting is approximately 40 mA. The transistor Q_2 has a medium current capability. The bias range is $+0.89$ to 1.32 V (for a 10-V supply). The circuit was tested with a 1.7-V LED. The driving circuit shown can be preceded by an audio amplifier or high-frequency pulse circuit.

LEDs or IREDs are designed to produce an efficient visual light or infrared output. Since plastic fiber-optic cables, as presently produced, have a minimum loss in the 660- to 820-nm wavelength, the LED must also operate in this range. The LED is effective if it is designed to produce a narrow beam of radiated energy in the same operating wavelength.

The LED's direct current can be modulated with an ac or pulsed signal. Capacity coupling or transistor driving circuits and preamplifiers can be used.

HIGH-POWER IRED LASER

Typical LEDs and IREDs have power outputs in the range of 100 to 900 μW and operate on a forward current of 10 to 50 mA. The transmitted power output is determined by the length of fiber cables to which it is coupled. A cable length of 1 to 5 m may require 50 mA. Figure 23-7 shows the typical electrooptical characteristics of a LED (SE4352 by Honeywell) used for coupling to a fiber cable. The peak output wavelength λp is 820 nm when the forward current is 50 mA.

Higher-power semiconductor laser diodes with peak emission in the range of 820 to 975 nm may have peak power outputs of 10 to 75 mW as compared to 100 to 900 μW for standard LEDs. (The limits indicated are examples only, and new technology will certainly extend their limits.) Although LEDs may respond to pulsed frequencies in the kilohertz range, some of the laser diodes will operate above 1 GHz. Figure 23-8 is a plot of a Hitachi laser. At a 50-mA forward current, the laser has a power output of 15 mW.

The laser is a quasiheterostructure in a domed structure. The laser structure is a mixture—Ga A1A—mounted on a silicon base. The radiated light output has a spectral response width, at the 50 percent power

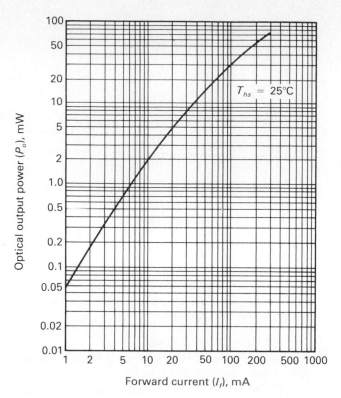

FIG. 23-8   Hitachi laser diode I_f versus output P_o power.

Parameter	Test condition	Symbol	Min.	Typical	Max.	Units
Forward drop	$I_f = 50$ mA	V_f		1.6	2.0	V
Series resistance		R_s		1.6		Ω
Device capacitance	$V_R = 1$ V	C_T		800		pF
Power output	$I_f = 50$ mA Aperture $= 1$ mm N.A. $= 0.5$					
—002			175	400		μW
—300			350	700		μW
Response time	1 V dc dias, 1 peak $= 10$ mA	t_f		12	20	ns
Peak output wavelength	$I_f = 50$ mA	λp		820		nm
Spectral bandwidth	$I_f = 50$ mA	Δd		35		nm
V_f temperature coefficient		$\Delta V_f / \Delta T$		-1.7		mV/°C
P_o temperature coefficient	$I_f = 50$ mA			-0.12		dB/°C
λ temperature coefficient		$\Delta \lambda / \Delta T$		0.35		nm/°C
Thermal resistance		θ		500		°C/W
Operating temperature			-55		85	°C

FIG. 23-7   Electrooptical characteristics of SE4352 ($T_c = 25$°C unless otherwise specified).

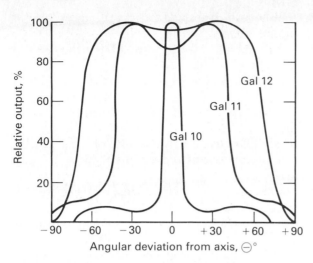

FIG. 23-9   Laser patterns (Plessey), infrared.

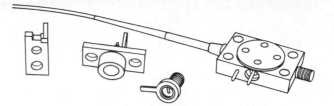

FIG. 23-10   Laser diode packing — Hitachi.

points, of 20 to 50 nm (bandwidth). Such diodes are used for interfacing to transmission cables. Other laser diodes, such as some produced by Plessey Opto-Electronics and Microwave Ltd., have broad spectral responses. Figure 23-9 shows the response of three diodes. The power output is 45 mW at 500 mA with a spectral frequency of 940 nm.

One application of a laser diode, such as the GaS12, is for individuals with hearing difficulties who attend the theater. When the auditorium is in darkness, infrared can be used for transmitting sound from the stage to those wearing special hearing aids equipped with infrared light sensors. A broad beam width is used.

A narrow beam is used in applications where fiber-optic cables are used in communications, printers, facsimile systems, video and audio transmission, table readers, interferometry, and spectroscopy. Figure 23-10 shows the packaging designs of some laser diodes from Hitachi.

High-power lasers require a metal casting for heat sinking, since high junction temperatures can destroy the diodes. The laser junction is also subject to damage from power supply transients; therefore, surge protectors are suggested for protection.

=========== **WARNING** ===========

THE OUTPUT POWER FROM A LASER IS INVISIBLE AND MAY BE HARMFUL TO THE RETINA OF THE HUMAN EYE. YOU SHOULD AVOID LOOKING INTO THE LASER, A CONNECTING CABLE, OR ANY OTHER DEVICE WHICH RADIATES SUCH ENERGY. A LASER BEAM IS USED TO PERFORM EYE SURGERY AND IS VERY EFFECTIVE IN CAUTERIZING BLOOD VESSELS. DO NOT LET THIS HAPPEN TO YOU ACCIDENTALLY!

TESTS AND MEASUREMENTS

MATERIALS REQUIRED FOR EXPERIMENT

Active Devices
SE4352 LED IRED
2N3904 Transistor, NPN
2N2222 Transistor, power, NPN

Resistors	Capacitors
5 Ω	0.1 μF (2)
10 Ω	0.22 μF
100 Ω	47 μF
120 Ω	100 μF
330 Ω	
470 Ω	
560 Ω	
1 kΩ	
2.2 kΩ	
3.3 kΩ	
6.8 kΩ	
18 kΩ	
1-kΩ Potentiometer	

The experimental tests relate to dc biasing of a LED and also driving the LED by using a transistor amplifier. Measurements will be made and validated.

1. Construct the circuit shown in Fig. 23-3. Use a 10-V dc supply. From the voltage drop across the resistor determine I_f.
2. What is the diode voltage?
3. Connect a 47-μF capacitor to the diode-resistor junction, and feed the capacitor with a 1-kHz signal. View the diode voltage on an oscilloscope. Adjust the level of the generator so that the output signal is not distorted. Find the peak-to-peak output voltage $V_{o(p\text{-}p)}$. Add half this voltage to the diode's voltage. What is the peak LED voltage?
4. Subtract the peak voltage from the 10-V supply to obtain the resistor's voltage drop V_R.
5. Determine the resistance of the LED (unmodulated).

109

AC Modulation

6. Construct the circuit shown in Fig. 23-6. Place a digital multimeter (DMM) across the 10-Ω resistor (TP_4), and adjust the dc control (1-kΩ potentiometer) so that the DMM reads 0.3 V. What is the current through the LED? Measure the voltage across the LED V_{LED}.

7. Set a function generator to 500 Hz, and feed a signal into the 47-μF capacitor. View the waveshape at TP_3. Increase the signal voltage until the waveshape starts to distort. Record the peak voltage across the LED V_{LED} and also that across the 10-Ω resistor, V_{10}.

8. The current increase due to modulation is $I_{mod} - I_c$. What are the values of I_g, I_{mod}, I_d, and I_F? Record the signal voltage at the generator output (signal-input terminal) V_g. Return the signal level to zero.

9. Determine whether a larger signal voltage is needed when the I_F current of the LED is increased. Set I_F to 40 mA, and again determine the maximum signal voltage across the LED. Is a higher signal voltage necessary?

10. Does modulation of the LED increase its light power output?

REVIEW QUESTIONS

On a separate piece of paper, complete the following statements with an appropriate word or words.

1. The light output of a LED is directly related to its _____ .

2. A LED can be dc-biased and _____ modulated at the same time.

3. The dc operating voltage of a LED is in the range of _____ V.

4. The LED is generally operated in the _____ and _____ spectrum.

5. The light spectrum of a LED must be matched to the _____ .

6. Most LEDs have a power output of _____ μW, while lasers have power outputs of above _____ mW.

7. For fiber-optic communications, lasers operate in the deep _____ region.

8. Infrared radiation from a laser is dangerous to the _____ .

9. If a LED is back-biased, you should be careful not to exceed the _____ rating.

10. A photometer is calibrated within a(n) _____ frequency range.

ANSWERS TO REVIEW QUESTIONS

1. Forward current
2. Pulse or ac
3. 1.2 to 1.7 V dc
4. Visual, Infrared
5. Cable and/or sensor
6. 100 to 900 μW, 2 mW
7. Infrared (820 to 975 nm)
8. Retina of the eyes
9. Reverse voltage
10. Specific

AM TRANSMISSION SYSTEM

SCOPE OF STUDY

This study covers amplitude modulation (AM) of a light-emitting diode. Also included in the study is the method of demodulating an AM signal.

OBJECTIVES

Upon completion of this study and experimental assignments and evaluation of the data, you will be able to:

1. Suggest applications for transmitting signals via an AM light-modulated system
2. Test and evaluate circuits similar to those presented in this experiment
3. Communicate your understanding of the advantages and disadvantages of an AM system to other technical people
4. Research LEDs for use in transmitters
5. Maintain and repair industrial/commercial equipment which incorporates light-modulated transmitters
6. Read and comprehend technical articles which describe the use of LED devices

BACKGROUND

In an amplitude-modulated radio system, an audio signal is used to modulate a radio-frequency (RF) wave operating at higher than audio frequencies. The RF wave serves as the carrier, and the audio signal provides the intelligence. In optoelectronic systems, light waves serve as the carrier. The light can be produced by either direct or alternating current. The intelligence can be conveyed in an audio, video, or pulse signal.

The audio section of a transmitter, in a communications system, consists of a transducer input, preamplifier, and audio driver feeding a modulator. The carrier section starts with an oscillator, driving an RF amplifier-buffer, and the output to the antenna is from an RF power amplifier. The carrier current, which produces electromagnetic waves, is varied (modulated) by the audio power amplifier in the RF power output stage.

The receiver in the system picks up the magnetic fields cutting across its antenna and converts them to small voltages. These voltages are processed in the mixer stage and IF amplifier. The original audio is separated from the carrier in the detector circuit, and the audio voltage is further amplified and then converted back to sound by the loudspeaker.

Figure 24-1 shows the circuit diagram for an AM light communications system. The transmitter (Fig. 24-1a) contains a two-stage audio amplifier (U_1 and U_2) and a light modulator (Q_1). The receiver input (Fig. 24-1b), a photodiode (in place of an antenna), feeds a two-stage amplifier (U_3 and U_4). The output of the amplifiers can be used to drive a speaker amplifier or other output device.

The amplifier U_1 in the transmitter has a stage gain of approximately 50 ($R_3 + R_4$). Resistor R_3 also serves as a gain control. The two-stage amplifier can provide full modulation when it is fed with a 1-mV signal. Amplifier U_2 also provides bias control for transistor Q_1. Resistors R_6, R_7, R_8, and R_9 establish the operating base current of Q_1 and the quiescent current of the LED.

The output of the transmitter is coupled to the receiver by means of a single-conductor fiber-optic cable. The gain of the amplifier U_3 varies with light changes on the photodiode. The greater the light (the lower its resistance), the higher the stage gain. Amplifier U_4 provides an additional stage gain of 5. A 1-mV signal to the transmitter will produce a receiver output of approximately 10-V peak to peak. The system has an overall gain of 10^4. In the experimental system, a signal is transmitted over a cable length of 0.5 to 3 m. The wavelengths of the photodiode and emitter are sensitive in the 820-nm spectrum. The cable is most sensitive from 660 to 770 nm.

The system can be used for high-linearity analog instrumentation, FM transmission, audio transmission, voltage-to-frequency conversion, and the multiplexing of data. In addition, the cable makes all signals immune to RFI and EMI.

ADDITIONAL STUDY—AN ALTERNATE AM TRANSMISSION SYSTEM

The circuit shown in Fig. 24-2 is a variation of Fig. 24-1a. In Fig. 24-2, a dynamic microphone with a built-in amplifier and gain control is used as the signal source. The microphone requires +7 V, which is obtained by using a 2N3904 as a zener diode. The IC amplifier (LM741) drives the current modulator.

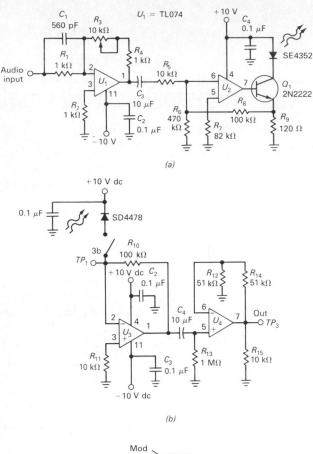

(a)

(b)

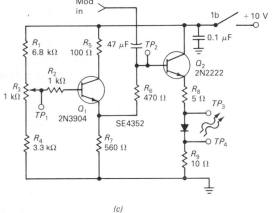

(c)

FIG. 24-1 Light communications system. (a) Transmitter. (b) Receiver. (c) Light driver (only).

In the receiver, a phototransistor feeds a 741C amplifier. A volume control determines the amount of audio signal which is fed to the power amplifier (LM386) and 8-Ω loudspeaker. Between the light emitter and pickup sensor is a fiber-optic cable which is terminated at both ends with AMP connectors. The circuit requires a ±10-V power supply.

The microphone shown is the type often used with citizens' band transceivers. All IC sockets (three) are eight-pin dual in-line. Care must be exercised in attach-

ing and aligning the emitter to its connector. The same applies to the phototransistor and its connector. This might best be done by feeding a low-level, 1-kHz, audio signal into pin 4 of the microphone jack. An oscilloscope should be connected across the loudspeaker. Improper alignment of the elements will result in no signal being transferred, or the signal voltage may appear clipped (half wave). The elements should be aligned to give a maximum reading on the oscilloscope.

If a microphone which does not have a preamplifier is used, an additional amplifier stage may be needed in the transmitter.

When the circuitry is properly functioning with an 8-Ω speaker and microphone, you should be able to speak into the microphone and hear the amplified voice from the speaker. Note that if the room light gets into the phototransistor, a humming tone will be heard (if artificial light is in use).

The power output of the LM386 is approximately 0.1 W (about the same as a small transistor radio).

TESTS AND MEASUREMENTS

MATERIALS REQUIRED FOR EXPERIMENT

Active Devices
SE4352 IRED, emitter
SD3443 Photodiode
TL074 (LF347) IC op amp
2N2222 Transistor, power
2N3904

Resistors		Capacitors	
5 Ω		0.1 μF (4)	
10 Ω		10 μF	
100 Ω	6.8 kΩ	47 μF	
120 Ω		560 pF	
1 kΩ (4)			
10 kΩ (2)	3.3 kΩ		
51 kΩ (2)	560 kΩ		
82 kΩ			
100 kΩ			
470 kΩ			
1 MΩ			
1-kΩ Potentiometer			
10-kΩ Potentiometer			
Fiber cable, 50 cm, with connectors			

In the experimental procedures, a transmitter, receiver, and a fiber-optic cable will be evaluated. Measurements

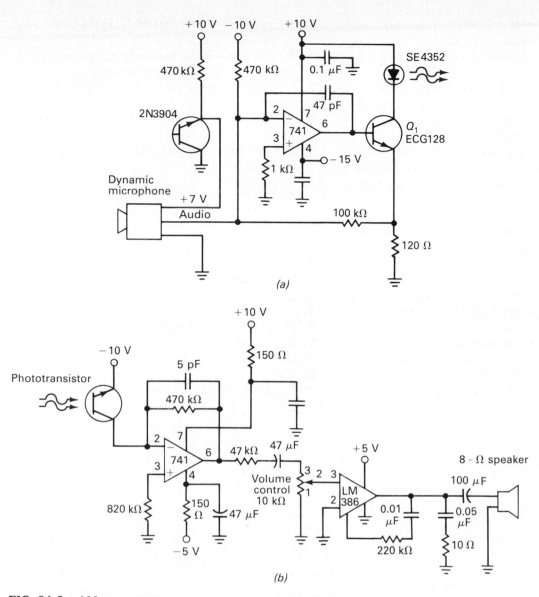

FIG. 24-2 **AM transmission system — optional. (a) Transmitter. (b) Receiver.**

will be taken to determine the operating characteristics of an AM system.

1. A transmitter will be coupled to a receiver via a fiber cable. For the transmitter, use the circuit shown in Fig. 24-1c. For the receiver, Fig. 24-1b is used. Both circuits operate on +10 V dc. A fiber cable (50 cm) is connected between the emitter and photodiode. (Close switches 1b and 3b.) Set the 1 kΩ potentiometer (R_3) on the transmitter for 300 mV at TP_4. Connect a function generator; set to a 100-Hz sine wave (zero voltage level) to the transmitter input (mod. in).

2. Connect one probe of a dual-trace oscilloscope to TP_3 (transmitter) and the other to the output of the receiver (also indicated as TP_3). Slowly increase the signal voltage until the receiver output reads

10 V peak to peak, or a lesser voltage if saturation or cutoff occurs. Record the receiver p-p output at TP_3. Record the transmitter output $V_{o(p-p)}$ at TP_3.

3. Measure signal input and output and compute the gain G of the receiver. Measure and sketch the waveform of the signal voltage at the receiver's output (TP_3).

4. Determine the frequency response of the transmitter and receiver while observing TP_3. Keep the generator constant, and vary the frequency from under 20 Hz to over 20 kHz. Where are the 3 dB points (0.707 times midband gain), at low F and high F?

5. What is the primary purpose of Q_2 in the transmitter?

6. How much gain is provided by U_4? What are the values of V_{in} and V_{out}?

7. From technical data available, determine the wavelength of radiation of the LED.
8. What is the peak wavelength of operation of the photodiode?
9. What is the operating dc voltage of the LED?
10. Determine the LED current flow I_F (dc) by measuring the voltage across the 10-Ω resistor.
11. What is the power input to the LED?
12. Feed the system with a square wave. What is the highest frequency the system will pass without severe distortion of the leading edge of the wave?

REVIEW QUESTIONS

On a separate piece of paper, complete the following statements with an appropriate word or words.

1. How could the transmitter be made more sensitive to ac signals? _____.
2. Where are system losses likely to occur? _____.
3. In the AM system demonstrated, the carrier is a(n) _____ current of the IRED.
4. As the IRED current I_F is decreased, the driving signal voltage must also be _____.

5. The gain of the first receiver stage is determined by the ratio of _____.
6. As the gain of the experimental receiver is increased, its frequency response is _____.
7. Is the AM link more sensitive to pulses than to sine waves? _____.
8. Can AM transmitters be overmodulated? _____.
9. The low-frequency response of the system is determined by _____.
10. A photodiode is generally _____ biased.

ANSWERS TO REVIEW QUESTIONS

1. Add audio stages before the IRED driver.
2. At the emitter, fiber cable coupling, light mismatch to components, and light input to the photodiode.
3. Direct
4. Decreased
5. R_{10} to the photodiode R
6. Decreased
7. No
8. Yes
9. Capacitors C_1 (47 μF) and C_7 (10 μF)
10. Back

AN FM CARRIER

SCOPE OF STUDY

This experiment covers the basic technique of frequency-modulating a carrier as well as the demodulation of the carrier at the receiver. A voltage-controlled integrated circuit is evaluated, and its frequency deviation and other characteristics are measured.

OBJECTIVES

Upon completion of the experimental assignments and evaluation of the data, you will be able to:

1. Describe the operation of an FM oscillator — light modulator
2. Suggest applications for FM transmission
3. Test and evaluate a basic FM transmitter and receiver which are the terminals of a fiber-optic cable
4. Communicate your understanding of the FM system to other technical people
5. Research optoelectronic components for specific applications
6. Maintain and repair industrial/commercial equipment using optodevices, given additional information and training on the overall operation of the equipment
7. Read and comprehend technical articles which describe the operation of FM light communication systems

BACKGROUND

Frequency modulation of a carrier wave has applications in telephone systems, instrumentation, and information transfer systems. Frequency modulation (FM) is used in frequency-division multiplexing (FDM) of telephone voices. The carrier wave, whose frequency may be over 25 kHz, is modulated in time. Three waves are present: the carrier, the modulating wave, and the modulated wave. The modulating wave can be low in frequency, such as the output from a pressure or temperature transducer or from an audio transducer. The modulating voltage shifts the frequency of the carrier in accordance with the amplitude of the modulating wave.

Frequency modulation is also referred to as *angular*, or *phase, modulation*. In frequency modulation the amplitude of the modulating wave determines the degree of frequency shift of the instantaneous wave from its unmodulated frequency. The magnitude of the frequency change, determined by the amplitude of the modulating signal, is called *frequency shift*, or *frequency swing*. Other terms relating to the frequency change are *frequency deviation* (maximum frequency change specified for the equipment) and *peak deviation*. The terms refer to a change in the carrier frequency caused by the amplitude change of the modulating wave.

In commercial broadcasting a maximum frequency deviation of ±75 kHz is permitted, and an additional 25 kHz is allotted on either side as a buffer or guard band between stations. Each FM station, therefore, has an allocated bandwidth of 200 kHz, of which 50 kHz cannot be used. This guard band is required because of the necessary bandwidth of the receiver's IF stages.

The modulating signal in FM causes a frequency change as compared with an amplitude change of the carrier in AM radio. During FM demodulation, approximately 25 percent of the amplitude is eliminated by clipping or limiting. This removes all forms of electric disturbances which may ride on the amplitude of the carrier with the information signal.

The rate of frequency deviation is related to the frequency of the modulating wave. If the modulating signal is 100 Hz when the carrier is 50 kHz, the carrier will shift at the rate of 100 Hz. The two important factors to be considered are *frequency deviation*, which relates to the amplitude of the modulating wave, and the *deviation rate*, which relates to the frequency of the modulating wave. Figure 25-1 shows the effects of a modulating sine wave signal on a carrier.

As the amplitude of the modulating voltage increases, the frequency change of the carrier increases. And when the modulating voltage decreases, the carrier frequency decreases.

An AM signal produces one upper and lower sideband, while an FM signal produces multiple upper and lower sidebands. For example, a 3-kHz signal modulat-

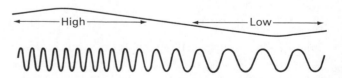

FIG. 25-1   Frequency-modulating a carrier.

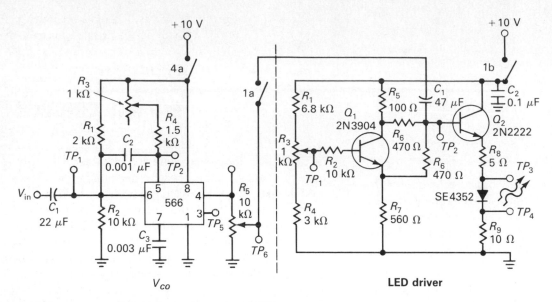

FIG. 25-2 Circuit diagram of an optical FM transmitter.

ing a 100-MHz carrier will produce multiple sidebands every 3 kHz. Five to seven such bands may appear on each side of the carrier frequency. The bandwidth may cover 40 to 50 kHz. With a maximum allowable deviation of ±75 kHz, a 15-kHz signal will produce a modulating index of 5 (75/15). FM produces a wider bandwidth than AM; however, this improves the signal fidelity. FM provides better fidelity than AM and at lower noise levels.

Frequency modulation can be produced by varying the frequency of the oscillator or by using the modulating signal to phase-shift the carrier. The latter is referred to as *phase-modulated* FM.

Frequency modulation can take place within the standard FM broadcast band, namely, 88 to 108 MHz, or at higher or lower frequencies. A carrier frequency of 10 to 100 kHz also can be easily frequency-modulated. In the laboratory experiment a voltage-controlled oscillator operating in the range of 50 to 80 kHz is frequency-modulated. Figure 25-2 shows a circuit diagram of an optical FM transmitter which can be used in a fiber-optic communications system. The transmitter has two main sections—the FM oscillator and the IRED driver. The carrier oscillator is a voltage-controlled device (VCD) with both a terminal and square wave output. A change in the voltage at either pin 5 or 6 will shift the oscillating frequency.

The frequency of oscillation can be computed from

$$F_o = \frac{2(V_{cc} - V_c)}{R_3 R_4 C_3 V_{cc}}$$

where V_{cc} is the supply voltage, V_c is the control voltage at pin 5, and R_3, R_4, and C_3 are the frequency-controlling elements.

The output signals of the oscillator are both a square wave (pin 3) and a triangular wave (pin 4). Other similar

devices available also produce sine waves. Figure 25-3 shows the relationship between the output frequency and changes in the dc voltage at pin 5 (TP_1). The voltage changes were made by varying R_3.

The second section of the transmitter is the IRED driver. The potentiometer R_3 determines the dc flow through the IRED. Figure 25-4 shows the relationship between the voltage level established by R_3 and the I_F current of the IRED.

A voltage change of 0.1 V produces a frequency change of approximately 3 to 4 kHz. As the voltage V_c goes higher, the frequency goes lower. An increase or decrease in control voltage can be accomplished by changing the setting of R_3 or by superimposing an ac voltage onto the dc voltage level. If an ac signal is low in frequency, the deviations of the output frequency can be observed on an oscilloscope. The voltage level of the

$+V_c$	F_o (kHz)
8.0	74.00
8.1	70.36
8.2	66.96
8.3	62.84
8.4	59.27
8.5	55.77
8.6	52.12
8.7	48.42
8.8	44.70
8.9	41.11
9.0	37.36
9.1	33.38
9.2	29.40
9.3	25.77
9.4	22.00

FIG. 25-3 Output frequency versus control voltage.

V_{R_3}	I_F (mA)
0.89	42
0.95	41
1.00	40
1.05	37
1.10	34
1.15	29
1.20	20
1.25	1
1.30	0

FIG. 25-4   IRED current versus drive voltage.

ac signal determines the deviation in frequency, and the frequency of the modulating signal determines the deviation rate and modulation index.

If R_3 is set to $+1.14$ V, the IRED current will be in the range of 30 mA. If the triangle signal voltage from the potentiometer R_5 is added to the IRED voltage, the IRED current will vary at the frequency of the FM oscillator. The IRED current is frequency-modulated. The modulated voltage across R_9 indicates the changes in the I_F current.

The position of the potentiometer R_5 determines the level of IRED driving current. If the ac voltage from the potentiometer is made too large, the IRED current will appear distorted. Transistor Q_1 is an amplifier while Q_2 determines the current flow of the IRED.

TESTS AND MEASUREMENTS

MATERIALS REQUIRED FOR EXPERIMENT

Active Devices	Capacitors
SE4352 Emitter	0.001 μF
566	0.003 μF
2N3904	0.1 μF
2N2222	22 μF
	47 μF

Resistors
5 Ω
10 Ω
100 Ω
470 Ω
560 Ω
1.5 kΩ
2 kΩ
3 kΩ
6.8 kΩ
10 kΩ (2)
1-kΩ Potentiometer (2)

The experiment relates to observing and evaluating the effects of modulating an IRED with a high-frequency ac voltage (carrier) and then frequency-modulating the carrier with a low-frequency signal.

1. The circuit shown in Fig. 25-2 will be evaluated. The circuit is to be constructed, or it may be provided, assembled and ready for evaluation.
2. Set the supply voltage to $+10$ V, and do not initially connect the function generator. (Close 4a and 1b.)
3. Observe the carrier current without 566 modulation. Connect the DMM to TP_3. Vary R_3 until the voltage at TP_3 equals approximately 1.8 V. Verify the current I_F through the IRED by measuring the voltage drop at TP_4 which is across R_9 (10 Ω), and compute the current I_F. Set I_f to 30 mA.
4. Close 1a and adjust R_5 for 0.3 V p-p at TP_4. Sketch the waveshape observed at TP_4 (emitter).
5. Sketch the waveshape at TP_3 (emitter).
6. As the IRED current (carrier current) is increased (smaller R_3 voltage), does distortion decrease? (If the carrier current is made too large a percentage of the dc IRED current, distortion occurs. Under 10 percent is desirable.)
7. Evaluate and record the frequency change of the oscillator (566) as a result of changing the control voltage on pin 5 (R_3). Connect a frequency counter and/or oscilloscope to TP_4 and a DMM to TP_2. Adjust R_3 so that the frequency at TP_4 is 55 kHz. From the values of R_3 and R_4 (estimated), C_3, and the voltage at pin 5 (V_c), compute the frequency F_o.
8. Observe the frequency change F_o by varying R_3. Compare your voltage and frequency readings with the values shown in Fig. 25-3.
9. Suppose an ac voltage were added to the dc voltage at pin 5. Would the voltage vary about the dc setting?
10. From the dc values, what is the frequency change per 0.1-V change in control voltage?
11. *Optional:* Plot the voltage and frequency of Fig. 25-3, or prepare your own table and plot. Is the frequency change a linear function of voltage?
12. Connect a function generator to the signal input V_{in}. Use a sine wave, 1 V p-p at 2 to 5 Hz. Observe the voltage at TP_2, TP_3, and TP_4. Synchronize the sweep, and note that the frequency varies. At the same time, observe the voltage swing at TP_1 or at the generator's output. Use an oscilloscope or an analog vacuum-tube voltmeter (VTVM). As the signal voltage increases (positive-going), does the output (TP_4) frequency increase or decrease?
13. In the discussion, we said that the larger the signal voltage, the greater the deviation. Increase the generator's voltage, and observe the frequency

change. Does an increased signal voltage increase the deviation of frequency?

14. Vary the frequency of the generator, and observe whether the deviation changes. Does the amount of deviation change, or does the rate change?

15. Is the voltage (waveshape) at TP_4 following the signal voltage? Observe the effect of square and triangular wave generator outputs.

16. Do your observations confirm the principles described in the discussion?

4. FM occupies a broader bandwidth than AM.

5. FM is more noise-free than AM.

6. The IC566 is a voltage-to-frequency converter.

7. The output voltage from the 566 is a sine wave.

8. The amplitude of the voltage from the 566 is determined by the power supply (V_{cc}).

9. Unlike AM, FM transmitters cannot be over-modulated.

10. The carrier frequency of an FM transmitter is generally crystal-controlled.

REVIEW QUESTIONS

Test your comprehension of the subject by answering the following questions. Answer True or False on a separate piece of paper.

1. The amplitude of the modulating signal determines the range of deviation.

2. In commercial broadcasting, a deviation of ± 75 kHz plus a band guard of ± 25 kHz is standard.

3. In FM the amplitude of the carrier remains constant.

ANSWERS TO REVIEW QUESTIONS

1. T
2. T
3. T
4. T
5. T
6. T
7. F
8. T
9. F
10. F

FM CARRIER DEMODULATION

SCOPE OF STUDY

The demodulator or receiver system is a continuation of the FM study presented in Experiment 25. In this experiment an FM receiver, capable of demodulating a frequency-modulated light beam, is evaluated.

OBJECTIVES

Upon completion of the experimental assignments and evaluation of the data, you will be able to:

1. Describe the operation of an FM light beam receiver
2. Suggest applications for FM principles
3. Test and evaluate circuits similar to those presented in this experiment
4. Communicate your understanding of the FM light beam receiver to other technical people
5. Research optoelectronic information as required for the understanding of an FM receiver
6. Maintain and repair industrial/commercial equipment using an FM transmitter/receiver link
7. Read and comprehend technical articles which describe variations in the use of FM separated links

BACKGROUND

An FM demodulator contains circuitry that enables a frequency-to-voltage conversion. Since EMI affects the amplitude of a signal, the demodulator generally includes some form of limiter which clips, clamps, or limits the amplitude of the incoming signal. Since the intelligence in an FM signal is within the carrier wave, such clipping does not hinder the reception or its fidelity.

In wireless telecommunications, the demodulator in the FM receiver takes the form of a discriminator, phase lock loop, ratio detector, or product detector circuit. Prior to one of these detector circuits in the receiver are the RF mixer and IF amplifiers.

Figure 26-1 shows a block diagram of a typical FM receiver. An FM light carrier receiver incorporates similar design concepts but different types of circuits. Figure 26-2 shows an FM carrier receiver, and the following briefly describes its operation.

The optosensor, receiving light energy from a fiber cable, converts the radiant light to an electric current. Whether the sensor is a photodiode or phototransistor will determine the number of amplifier stages that will follow. In the experimental circuit, two stages of gain increase the carrier voltage to such amplitude that its voltage overdrives the limiter amplifier. When the amplifier is overdriven, the upper and lower portions of the carrier are clipped (wave peaks are flattened). Any noise riding on the peaks of the carrier is removed, and the output waveshape resembles a square wave. Since the intelligence is within the carrier, clipping does not affect the signal quality.

The demodulator is a frequency-to-voltage converter. The output of the demodulator is an audio signal which can be further amplified by an audio amplifier. Figure 26-3 shows the circuit diagram for the laboratory evaluation project. The receiver can handle a carrier frequency of 50 to 100 kHz. The filter, formed by C_{17}, L_1, C_8, and R_{14}, rejects the carrier in the output.

Sensor Q_1 is a phototransistor whose gain is controlled by feedback from Q_2. Transistors Q_2 and Q_3 provide two stages of gain, and Q_4 is the limiter. Transistor Q_5 is a silicon unilateral switch (SUS). This transistor is a four-layer device which is similar to a triggered zener. The device operates from 6 to 10 V. During demodulation, the square wave (clamped) carrier provides 7 V at the anode and 8 V at the gate.

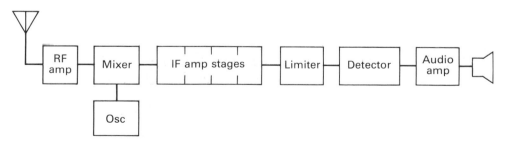

FIG. 26-1   Block diagram of an FM receiver.

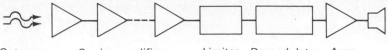

Optosensor   Carrier amplifiers   Limiter   Demodulator   Amp

FIG. 26-2   FM carrier receiver.

Transistor Q_6 has a square wave on its base and variations in pulse on its emitter. The variations are due to the charge and discharge of capacitor C_6. The SUS clamps the voltage at 7 to 7.5 V. Negative-going pulses from Q_4 turn on the SUS. The collector output of Q_6 contains the audio signal with a residual carrier. Capacitors C_7 and C_8 and coil L_1 make up a pi filter which eliminates the remainder of the carrier. Capacitive effects at the sensor and filtering at the output limit the audio frequency range to under 20 kHz. The type of sensor used will be determined, in part, by the operating frequency of the IRED and the desired gain of the receiver.

TESTS AND MEASUREMENTS

MATERIALS REQUIRED FOR EXPERIMENT

Active Devices
MFOD200 Phototransistor
2N4987/ECG6404 Silicon unilateral switch
2N3904 (3) Transistors, NPN
2N3905 Transistors, PNP
IN4148 Silicon diode

Resistors	Capacitors
10 Ω	100 pF
100 Ω	200 pF
470 Ω	390 pF
1 kΩ (2)	0.002 μF
10 kΩ (2)	0.01 μF (2)
15 kΩ	0.03 μF
47 kΩ (4)	0.1 μF
330 kΩ	47 pF
4.7 MΩ	0.2 μF
22 MΩ	

Miscellaneous
10-mH Inductor
50 cm of fiber cable (OE1040) with
    optoconnectors
3 m of fiber cable (OE1040) with optoconnectors
Transmitter from Experiment 25

In the laboratory an FM carrier receiver will be evaluated. Waveshapes will be observed and measurements

made. A complete system comprising a transmitter and receiver will be tested.

1. Construct and/or connect the circuit shown in Fig. 26-3. The circuitry requires 10 V. The circuit also requires an FM transmitting signal source. Make the appropriate connections to operate the transmitter circuit shown in Fig. 25-2. Between the transmitter and receiver use a 50-cm fiber-optic cable with appropriate connectors.

2. Properly adjust the transmitter by connecting an oscilloscope to TP_3, so as to observe the IRED's carrier modulating wave. Adjust R_5 and R_3 in the modulator until a clearly defined triangular waveshape is observed at the IRED output.

3. Close switch 2b. Observe the carrier waveshape and voltage in the receiver at TP_7, TP_8, and TP_9. Carefully adjust R_5 or R_3 in the transmitter since the photosensor may overload for short cable lengths. Sketch the waveshape, and record the peak-to-peak voltages at all three points.

4. Stray capacitance at the input of the receiver affects the signal waveshape. What is the capacitive effect on the triangular wave?

5. From the measurements made, compute the stage gain G of Q_2 and Q_3 and the total gain.

6. Observe and record the waveshape of the limiter (Q_4) at TP_{10}. Use the peak-to-peak output versus the input for the gain measurements. What are V_o at TP_{10} and V_{in} at TP_9? What is the gain?

7. Is Q_4 providing a limiter action?

8. Is the modulating frequency still the same as at the transmitter's carrier?

9. The SUS operates in the switching range of 6 to 10 V. How much switching voltage is present at TP_{12}?

10. Record the signal at TP_{11}.

11. The output of Q_6 connects to a 50-kHz pi filter. Observe and record the wave at TP_{13} and TP_{14}. Is the carrier wave attenuated by the filter?

12. Use your oscilloscope or a counter to determine the carrier frequency.

13. Connect a function generator to the transmitter, using a 0.2-V 500-Hz square wave signal. Observe the waveshape at TP_{14} while increasing the amplitude of the generator. Readjust all controls carefully since overdriving the system will distort the output. Sketch the output (TP_{14}) waveshape and

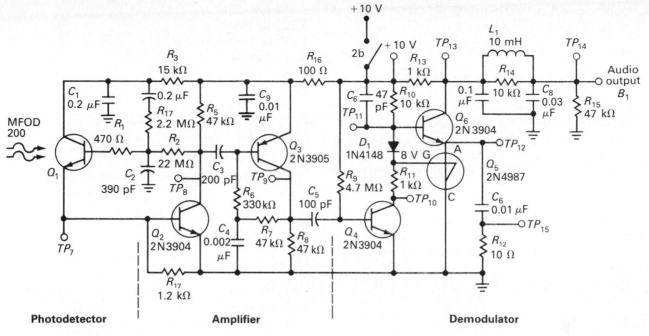

FIG. 26-3 Carrier receiver, 50 kHz.

signal amplitude at 500 Hz. The sine and triangular waves should also be reproduced.

14. Set the function generator to a minimum signal level and view TP_{12}. Only the carrier square wave should be present. Slowly increase the amplitude of the generator. What happens to the square wave at TP_{12}?

15. The frequency deviation of an FM signal depends on the amplitude of the modulating signal. Does the carrier frequency shift increase with a greater signal amplitude?

16. The capacitor C_6 (0.01 μF) and R_{12} (10 Ω) form a differentiator. What is its waveshape at TP_{15}? Record the time period between spikes ($T = 1/F$).

17. The SUS is a triggered zener. When the positive pulse exceeds the breakdown of the SUS, the gate is triggered on and the SUS discharges. Positive pulses at TP_{15} result from the rising square wave, and negative-going pulses are derived from the negative slope of the square wave. Is the time between the rising (positive) and falling pulse at TP_{10} the same as the time of the signal?

18. Modulate the transmitter, using a frequency between 3 and 10 Hz. Slowly increase the amplitude of the generator. Describe what happens at TP_{14}.

19. Determine the frequency response of the system between the 3-dB points.

20. Modulate the carrier signal with a 1-kHz 0.2-V-amplitude signal, and observe its waveform at both TP_{13} and TP_{14}. Does the pi filter remove the carrier signal?

21. Is the system suitable for telephony? Why or why not?

22. If the phototransistor were changed to a photodiode, would more or less amplification be needed in the receiver?

23. Does a phototransistor normally have a higher or lower frequency response than a photodiode?

REVIEW QUESTIONS

On a separate piece of paper, complete the following statements with the appropriate word or words.

1. The detector in an FM receiver is a(n) _____ converter.

2. The sensitivity of an FM (light) receiver is controlled by the _____ and the overall _____.

3. The audio response of the receiver is determined by the _____ and the output _____.

4. A SUS device is a form of _____.

5. Is the MFOD200 a photodiode, phototransistor, Darlington, or SUS? _____.

6. In the experimental circuit, the gain of the sensor is controlled by resistors _____.

7. The inductor in the output of the receiver circuit presents a high impedance to the _____.

8. The SUS has an operating voltage of approximately _____ V.

9. The higher the carrier frequency, the narrower will be the receiver's _____.

10. In the receiver circuit, transistor Q_4 serves as a(n) ———— .

ANSWERS TO REVIEW QUESTIONS

1. Frequency-to-voltage
2. Sensor, stage gain
3. Sensor, filter
4. Zener diode
5. Phototransistor
6. R_1 and R_2
7. Carrier
8. 7
9. Bandwidth
10. Limiter

PULSE-WIDTH MODULATION

SCOPE OF STUDY

This experiment as well as Experiment 28 covers the methods of pulse-width modulation of a signal, the transmission of the signal over a fiber-optic line, and the demodulation of the signal at the receiver. An astable-monostable circuit provides the carrier frequency which is pulse-width-modulated.

OBJECTIVES

Upon completion of the experimental assignments and evaluation of the data, you will be able to:

1. Describe the principles and techniques used to produce pulse-width modulation (PWM)
2. Suggest applications for the principles involved in PWM
3. Test and evaluate transmitter circuits similar to those presented here
4. Communicate your understanding of PWM techniques to other technical people
5. Research optoelectronic components for use in PWM transmitters
6. Maintain and repair industrial/commercial PWM equipment using optodevices, given additional information and training on the overall operation of the equipment
7. Read and comprehend technical articles which describe the use of PWM techniques

BACKGROUND

Pulse-width modulation, like AM and FM, can be used for the transmission of low-frequency signals. In PWM, the amplitude and frequency of the carrier are held constant while the pulse width varies. The carrier is generally a square wave (pulses), and the modulating signal varies the width of the pulses. A modulator converts analog signals to a sequence of pulses with different pulse widths. The pulse-width variation contains the intelligence supplied by the modulating signal.

Pulse-width modulation is also referred to as a pulse-time modulation (PTM), pulse-position modulation (PPM), and pulse-duration modulation (PDM). The modulating signal may vary the leading or trailing edge of the carrier pulses or both. In pulse-position modulation, the modulated pulses are shifted in time from their original location. Figure 27-1 shows the general appearance of several forms of modulation.

The transmission of modulated pulses, rather than analog signals, reduces noise and distortion levels. Unlike analog signals, pulses are easily reproduced and can be transferred through repeater stations. Although noise or interference may be present, the transmitted carrier pulses, which may be deteriorated, can be used to retrigger a fresh chain of similar pulses. The source of analog voltage can be a voice or an instrumentation transducer.

The transmitter consists of an oscillator (square wave) and a pulse generator (Schmitt trigger or monostable oscillator). The modulating signal varies the pulsewidth of the pulse generator while the repetition rate is determined by the oscillator.

If the control voltage of the oscillator is modulated, the oscillator becomes a pulse-position modulator.

In the laboratory circuit, a dual timer (LM556) is used for the oscillator and pulse generator. Figure 27-2 shows the pin layout for this integrated circuit. Figure 27-2a is used as the pulse generator and Fig. 27-2b as the trigger oscillator.

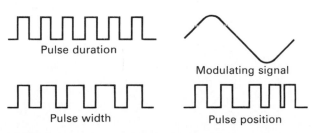

Pulse duration

Modulating signal

Pulse width

Pulse position

Pulse amplitude

FIG. 27-1   Forms of modulation.

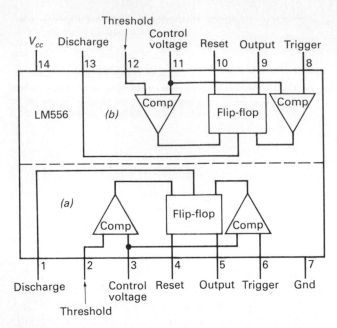

FIG. 27-2   Dual timer — LM556.

The total period is given by

$$T = t_1 + t_2 = 0.693(R_6 + 2R_7)C_6$$

and the frequency of oscillation is given by

$$f_o = \frac{1}{T} = \frac{1.44}{(R_A + 2R_B)C}$$

The duty cycle is given by

$$D = \frac{R_B}{R_A} + 2R_B$$

The oscillator should have a frequency which is lower than the frequency of the pulse generator: $f_o = 1/t_1$.

Both oscillators are triggered by negative-going pulses. The output from the astable oscillator (TP_3) is fed to the trigger of the monostable (pin 6). Diode D_1 clips the positive-going pulses.

Components R_1 and C_1 determine the period of the monostable pulse generator. For the component values shown, the pulse width is approximately 100 μs. The modulated pulse output is observed at pin 5, and the modulating input signal is fed into pin 3. Figure 27-4 shows the pulse-width waveshapes produced when the circuit is modulated by a sine wave.

During the laboratory experimentation, a PWM circuit will be modulated and the patterns observed. The output will be used to feed a LED/IRED driver circuit.

Figure 27-3 shows the complete circuit of the PWM transmitter. The frequency of the oscillator (Fig. 27-3b) is established by components $R_6 + R_7$ and C_6. Resistors R_6 and R_7 determine the charging time of C_6 (period t_1). Components R_7 and C_6 determine the discharge period $t_2 = 0.693R_7C_6$. The charge time (high output) is given by $t_1 = 0.693(R_6 + R_7)C_6$.

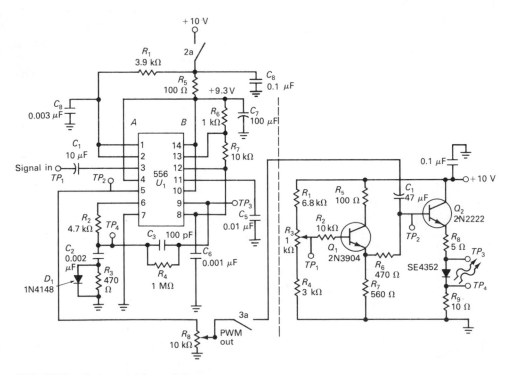

FIG. 27-3   Pulse-width modulator.

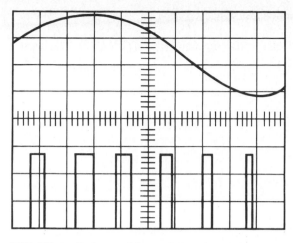

FIG. 27-4 Pulse-width-modulated pattern, using a sine wave signal.

TESTS AND MEASUREMENTS

MATERIALS REQUIRED FOR EXPERIMENT

Active Devices
SE4352 IRED
556 Dual timer
2N3904 NPN transistor
2N2222 NPN power transistor (Q_2)
IN4148 Diode

Resistors	Capacitors
5 Ω	100 pF
10 Ω	0.001 μF
100 Ω	0.002 μF
470 Ω (2)	0.003 μF
560 Ω	0.01 μF
1 kΩ	10 μF
3 kΩ	47 μF
4.7 kΩ	100 μF
6.8 kΩ	
10 kΩ	
1-kΩ Potentiometer	
10-kΩ Potentiometer	

In the experiment you will measure and record the frequency of the oscillator and pulse generator, modulate the pulse generator to produce pulse-width modulation of an IRED, and modulate the astable oscillator to produce pulse-position modulation.

1. Arrange the circuit shown in Fig. 27-3. Use a +10-V supply. Observe and record the oscillator's output at TP_3, using an oscilloscope and/or fre-quency counter to measure the oscillator's fre-quency output. Find t_1, t_2, f_o^2, and $V_{op \cdot p}$.

2. View and record the pulse generator's peak-to-peak output signal at TP_2. Find the pulse width and duty cycle.

3. View the pulse output at TP_2 during modulation. Use a 1.0-V p-p sine wave of 500 Hz as the modulat-ing signal.

NOTE: Pulse-width modulation produces horizontal jit-ter. Sketch the unmodulated and modulated output waveforms.

4. What happens to the waveshape of the duty cycle observed at TP_2 when a modulating signal is applied to TP_1?

5. Observe the effects of overdriving (overmodulating) the oscillator.

6. Adjust the PWM control R_8 for a zero output signal. Adjust the IRED driver (R_3) until the direct-current voltage (minus any ac present) at TP_4 (R_9) equals 0.3 V. What is the I_f current?

7. Increase R_8 until a peak signal of 3 V p-p appears at the driver output (TP_4). Record the voltage and waveshape at TP_4. How much current flows through the IRED when the signal is present? Has the output increased during modulation?

REVIEW QUESTIONS

On a separate piece of paper, complete the following statements with the appropriate word or words.

1. Would electromagnetic noises affect the PWM? Why or why not? _____

2. From the noise reduction observed, PWM is simi-lar to _____ .

3. The oscillator in the circuit must have an operating frequency which is _____ than the pulse genera-tor frequency.

4. In PWM, the carrier frequency is set by the _____ .

5. The oscillator's output pulse amplitude is deter-mined by the _____ .

6. The carrier frequency is a(n) _____ wave.

7. The amplitude of the modulating signal deter-mines the _____ . The higher the amplitude, the _____ the pulses.

8. Overmodulation cannot occur in pulse-width mod-ulation if the _____ is made too large.

9. The frequency of the carrier can be changed by altering components _____ .

10. The power output of a PWM transmitter would be determined by the _____ and the _____ of the modulator.

ANSWERS TO REVIEW QUESTIONS

1. No. The noise would ride on top of the modulated pulses and be clipped off.
2. FM
3. Greater
4. Oscillator
5. Power supply
6. Square
7. Pulse width, narrower
8. Signal
9. R_6, R_7, or C_6
10. IRED power, pulse amplitude

PULSE-WIDTH DEMODULATION RECEIVER

SCOPE OF STUDY

The PWM receiver is used in conjunction with the PWM transmitter. The receiver contains a photodiode sensor, active filter, and audio amplifier. The receiver's performance is evaluated during the demodulation of a signal.

OBJECTIVES

Upon completion of the experimental assignments and evaluation of the data, you will be able to:

1. Evaluate pulse-width-modulated receivers, given adequate data
2. Evaluate pulse-width-modulated transmission links
3. Evaluate a narrow-bandpass notch filter

BACKGROUND

The principles of pulse-width modulation were discussed in Experiment 27. In this experiment, a receiver

designed to demodulate a pulse-width signal is evaluated. The receiver circuit is shown in Fig. 28-1.

The pulse-width signal consists of unidirectional pulses which are generated from a carrier oscillator. The oscillator can be free-running or crystal-controlled. For experimental purposes an astable oscillator with a frequency of 50 to 60 kHz is used. The higher the oscillator frequency used, the higher can be the modulating frequency; but the bandpass of the receiver and cable link must also be wider.

The pulse width of the transmitted pulses is determined by the amplitude of the modulating voltage. The pulses can be varied until the modulating signal causes the pulse width to extend to the adjacent pulse. At this point overmodulation takes place, and the audio signal breaks up. Once again, the narrower the pulses, the higher the frequency response required by the system.

In this experiment a complete simplex communication system (one-way), transmitter and receiver, is established. The overall transmission system consists of a transmitter containing the following:

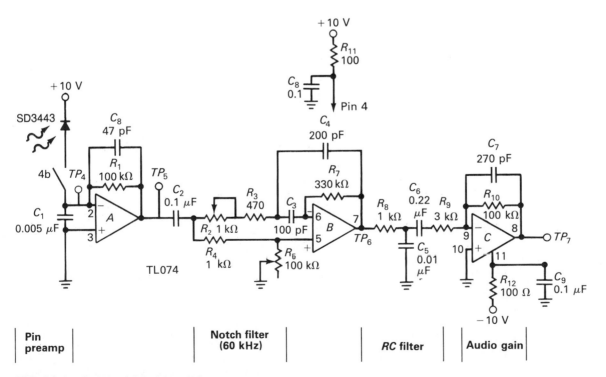

FIG. 28-1   Pulse-width demodulation receiver.

1. Carrier oscillator (output at TP_3 of the IC556, Fig. 27-3)
2. Pulse generator (output at TP_2)
3. Modulating input signal (fed to TP_1)
4. Light (IRED) driver

and a receiver with

1. Photodiode detector and amplifier A
2. Carrier filter, amplifier B
3. Audio amplifier C

The transmitter and receiver are connected by a fiber-optic cable. The greater the cable length (and cable loss), the greater the gain required in the receiver and the greater the driving power of the IRED.

In the receiver circuit shown in Fig. 28-1 a photodiode is used, since it provides a wide bandpass. The following describes the function of the components used in the receiver. Capacitor C_1 reduces the amplitude of the carrier frequency after detection by the photosensor. Amplifier A provides stage gain which is controlled by R_1. The amplified signal is fed to an active notch filter which is tuned to the carrier frequency by potentiometers R_2 (1 kΩ) and R_5 (100 kΩ). The two controls are adjusted for a minimum carrier signal at the output of amplifier B. The filter provides high attenuation (over 40 dB) at the tuned frequency.

The output of the notch filter feeds a low-pass filter, R_8-C_5, which further attenuates the carrier frequency. Amplifier C provides an additional stage gain for amplifying the audio-modulating signal voltage. Capacitor C_7 also provides carrier filtering.

The amplifiers are part of a quad IC, TL074 (similar to LF347 or MO3403), which has an RC carrier filter in each of the positive (R_{11}-C_8) and negative (R_{12}-C_9) power supply lines. Note that the astable oscillator and pulse generator consume 30 to 40 mA, and spikes can easily circulate in the supply or ground lines.

During the experiment a PWM communications system will be evaluated. The system can be used to transmit tone signals, speech, or other data. The system has a low-frequency bandpass of greater than 10 kHz.

TESTS AND MEASUREMENTS

MATERIALS REQUIRED FOR EXPERIMENT — RECEIVER ONLY

Active Devices
TL074 IC BIFET quad
SD3443 Photodiode

Resistors	Capacitors	
100 Ω (2)	47 pF	0.01 μF
470 Ω	100 pF	0.1 μF (3)
1 kΩ (2)	220 pF	0.220 μF
3.1 kΩ	0.0047 μF	
100 kΩ (2)		
330 kΩ	**Cable**	
1-kΩ Potentiometer	50 cm OE1040	
100-kΩ Potentiometer	Transmitter CKT (Fig. 27-3)	

In this experiment you will modulate the transmitter that was used in Experiment 27, using a sine wave from a function generator, and observe this signal at various test points in the system.

1. Arrange the receiver system shown in Fig. 28-1, using ±10-V supplies. Close switch 4b (all others off). Connect a 50-cm fiber cable between the IRED emitter of the driver and the photodiode. Set the function generator to 1 V p-p at 1 kHz, and feed this signal into the transmitter's input (TP_1 of PWM, Fig. 27-3). Review Experiment 27 on adjusting the transmitter. The IRED driver (emitter) current should be set for 25 to 30 mA. Check the waveshape voltage at TP_3 of the transmitter to make sure that a signal is being transmitted.
2. View the modulated output of the receiver at TP_5. Sketch the waveshape seen at TP_5.
3. View and record the output signal at TP_7. If the waveshape observed contains the carrier frequency, adjust both R_2 and R_5 to reject the carrier signal seen in the output. If jitter is present, vary the function generator's output to eliminate same. From the voltages present at TP_5 and TP_7, determine the voltage gain of the last two stages, B and C.
4. Record the carrier frequency (transmitter's oscillator), using your oscilloscope and/or a frequency counter. What is the value, in kilohertz, of TP_3 at 556?
5. What is the system bandpass between the 3-dB low and high points of the overall link (TP_4 of the transmitter to TP_7 of the receiver)?
6. Check the tuning of the notch filter by disconnecting the fiber-optic cable. Feed TP_4 of the receiver with a low-level generator signal set to 1 kHz. Record the output of the filter (TP_6). Do not overdrive the filter. What is the gain (combined with amplifier A) of the filter? Measure the voltage output at the notch frequency. Note that frequency adjustment is extremely critical. Compute the attenuation of the

filter in decibels. (Note that the filter responds to harmonics of the desired frequency.)

7. The bandwidth of the notch is quite narrow. Measure the width of the notch at the halfway (6-dB) points.

REVIEW QUESTIONS

On a separate piece of paper, complete the following statements with the appropriate word or words.

1. As the fiber cable is increased in length, what will be the effect on the carrier pulses? _____ .
2. Can a PWM system be overmodulated? _____ .
3. In a PWM system the amplitude of the pulses _____ vary with modulation.
4. The gain of the receiver can be increased by increasing _____ , _____ , _____ , _____ .
5. How much is the voltage gain of the overall receiver system (from the signal input voltage at TP_1 of the transmitter to TP_7 of the receiver)? _____ .
6. If the modulating signal were a temperature change, would it be received at the receiver output? _____ .

7. Could PWM be used as a form of voice communication? _____ .
8. Increasing the carrier frequency will _____ the overall bandwidth of the system.
9. A phototransistor would _____ the gain of the receiver but _____ the bandpass.
10. Amplifier C in the receiver can be increased by the resistor ratio _____ .

ANSWERS TO REVIEW QUESTIONS

1. They are likely to stretch and the corners will roll off.
2. Yes
3. Does not
4. R_1, R_7, R_9, R_{10} or changing the photodiode to a phototransistor
5. 12
6. Not likely. The receiver is ac-coupled, hence it has a low-frequency cutoff. Temperature changes are likely to be under 1 Hz.
7. Yes
8. Increase
9. Increase, decrease
10. R_{10}/R_9

MULTIPLEX TRANSMISSION

SCOPE OF STUDY

The multiplexing of signals enables numerous sources of information to be transmitted by means of a single metal cable, an RF carrier, a microwave carrier, or a light source modulating a fiber-optic link. Each signal of intelligence is on the carrier for only microseconds.

OBJECTIVES

Upon completion of the experiment and evaluation of the data, you will be able to:

1. Describe how several signal sources can be sampled, transmitted, and reassembled at the receiver
2. Test and evaluate a basic multiplexing system which can handle the transmission of several signals on a single light beam carrier
3. Generally understand how human voices can be digitally coded, multiplexed, and transmitted over a fiber-optic link

BACKGROUND

Time-division multiplexing is the most direct and basically the simplest method of multiplexing which is in common use. Each signal to be transmitted is successively connected to the transmission media for a very short time. In earlier times, a motor-driven commutator was used for connection to each signal source. Pres-

ently multiplexing is done by integrated circuits. Figure 29-1 shows the basic concept of a switching system. The switching can be a motor-driven switch or an IC circuit. Obviously, the latter is now in use.

The signal from each circuit is first converted to a series of pulse samples. During the sample period, the signal is transmitted. The signal can be dc, ac, AM, FM, PWM, PCM, or any other type of signal. The sampling rate is not critical. However, the rate determines the bandpass of the system, and it is important that the sampling rate be at least twice the frequency of the highest modulated wave to be transmitted. For voice transmission the sampling rate of telephone lines is usually 8000 Hz or higher. The normal bandwidth for telephone lines carrying voices is 300 to 3400 Hz. A 600-Hz guard is provided between adjacent channels. At an 8000-Hz clock rate, each frame of voices (24 slots) is sampled for 125 μs. The 24 slots each occupy 5.2 μs, and each is a voice channel. Slots of less than 5.2 μs can also be used to transmit audio signals. As more channels are added, the bandwidth has to be expanded in order to pass the fundamental sample rate and its harmonics.

In the laboratory project a sampling rate of approximately 400 kHz is used, and the bandwidth of the system is over 1 MHz. At narrower bandwidths, the transmitted pulses roll off, and the data sample becomes distorted.

The synchronization of the transmitting and receiving channel is critical. When an information signal is

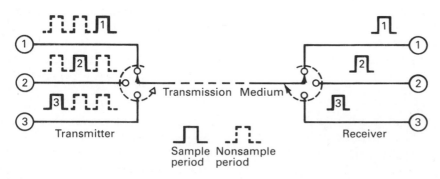

FIG. 29-1 **Basic multiplexing.**

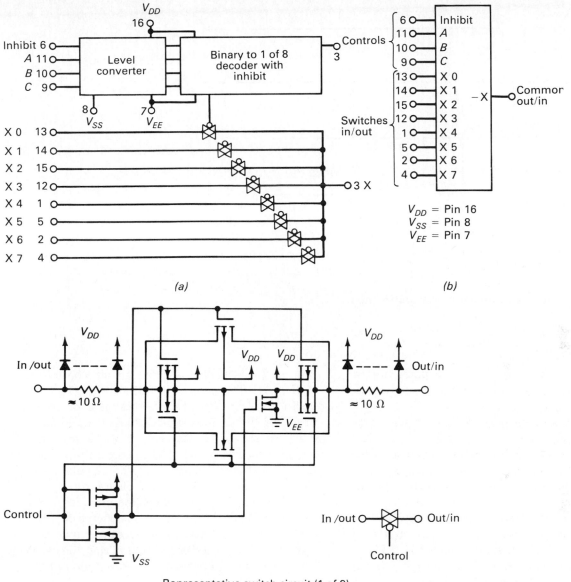

(a)

(b)

V_{DD} = Pin 16
V_{SS} = Pin 8
V_{EE} = Pin 7

Representative switch circuit (1 of 8)

(c)

FIG. 29-2 MC14051B multiplexer.

connected to channel 1, its receiver must also be on channel 1. In time-division multiplexing, a means of synchronizing must be provided. Generally, a *marker* pulse is transmitted which is different from the other data being transferred. The marker pulse can, for example, be greater in amplitude or wider. It is not important what shape is used, only that it can be recognized by the receiver and be used to synchronize the receiver with the transmitter.

A set of time slots, one for each channel, is referred to as a *frame*. The synchronized pulse can arrive at the receiver at the beginning or end of a frame. In this experiment, the marker pulse has a greater amplitude

than the channel pulses, hence it is recognized by the receiver and used to lock in the receiver's local oscillator. This synchronization pulse is located in the fifth time slot.

In this experiment IC analog switches, also called *commutational* switches, are used. Such IC switches are arranged in sets of 4, 8, or 16. The inputs and outputs of the IC can be isolated or grouped together. Some chips have multiple inputs and a single output, while others may have a single input and multiple outputs.

Figure 29-2a shows an IC with eight inputs and one output. Figure 29-2b (a subdetail drawing of Fig. 29-2a) shows the chip wiring. The circuitry uses CMOS tech-

131

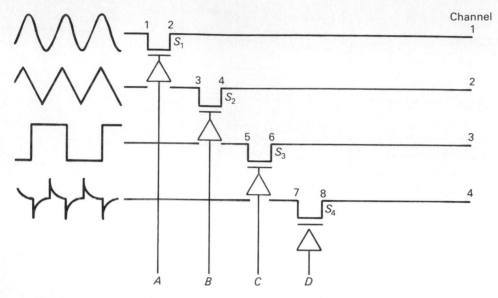

FIG. 29-3   Four-switch IC with gates and control inputs.

nology. Figure 29-2a shows how the gates are sequenced on with a logic sequencer connected to the control input (A, B, C).

Each switch is a FET, and some ICs include a gate which can be triggered on by a logic pulse. Figure 29-3 shows an IC with four switches and logic inverters. Four different signal sources are sequenced on. The four waveshapes shown are produced by a function generator which is a part of the experiment.

When input A is triggered on, the gate of switch S_1 is made positive, and the resistance between terminals 1 and 2 drops from a high resistance value of 10^{12} Ω to a resistance of under 100 Ω. When the switch closes, a signal voltage is passed from terminal 1 to terminal 2. The signal to be transmitted is on for the duration of the pulse period applied to the gate.

Each gate in the multiplex is driven on in succession by a ripple counter, timed by an oscillator, so that the timed square wave pulses ripple through each gate, A, B, C, and D. The sample rate is established by the clock frequency of the oscillator. The operation of a multiplexer system is explained by using the experimental circuitry as an example.

In this experimental circuit, an MC14051B analog multiplexer (U_1), which is digitally controlled, is used for sampling analog signals at the input to the transmitter and IRED driver. The 14051 IC is a single-pole, eight-position electronic switch. In the experimental circuit, some of the unused switches have been grounded. Figure 29-2b shows the pin layout of the chip.

The multiplexer has eight inputs which are accessed through pins, 13, 14, 15, 12, 1, 5, 2, and 4. The output from the chip is taken from pin 3. The logic sampling

rate for each channel is provided via control inputs A, B, and C. The chip can handle switching frequencies up to 65 MHz. When a channel (gate or switch) is turned on, it has a resistance of less than 60 Ω.

Figure 29-4 shows the circuit of the experimental multiplexing transmitter. The U_1 chip is the analog multiplexer with eight inputs. Only five inputs are used. Note that pin 1 in the circuit diagram is connected directly to pin 16. This connection provides +5 V on channel 5. When the digital sampling turns on this gate, a 5-V fast rising pulse is produced at the output, and this positive pulse is used for synchronizing the receiver. This is explained further in the description of the receiver circuit.

The four inputs of the transmitter are marked channels 1 to 4, and they can be supplied with analog or pulse signals whose amplitude is within the 5-V synchronizing pulse amplitude. The composite output at pin 3 feeds the light driver. The light-modulated signal is a composite of all channels.

The IC marked U_2 is a 7490 binary counter. This IC provides the digital sampling pulses through the control inputs of U_1. And U_2 is a logic divider which controls the three inputs (A, B, C) of the IC 14051. The binary counter is wired so that it only provides a count to 5. This enables four analog switches plus the synchronizing channel to be turned on in sequence. On the start of the sixth count, U_2 recycles to the first count.

The sampling rate of U_2 is controlled by a clock (U_3) which uses an LM555 timer. The clock frequency is approximately 400 kHz. The frequency of the clock is not critical, but as previously mentioned, it does control the bandwidth of the signals that can be passed through

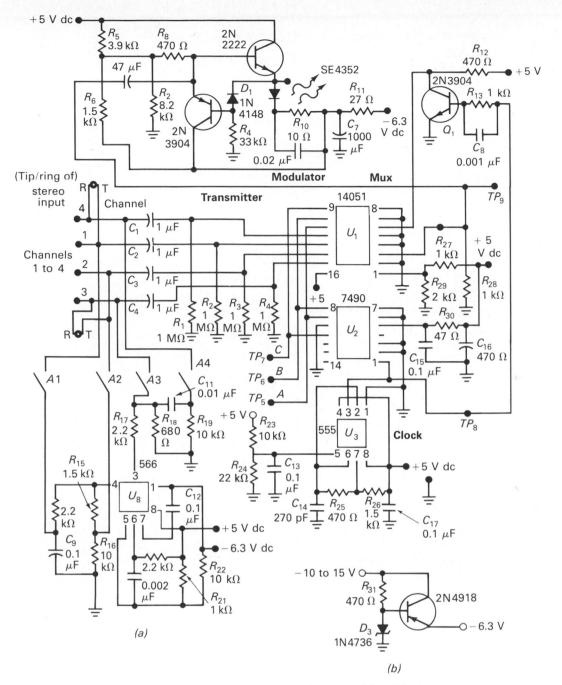

FIG. 29-4 Multiplexing transmitter. (a) Function generator. (b) −6.3-V supply.

the analog switches. The following is a summary of how the transmitter operates.

A clock (U_3) determines the basic operating frequency used for controlling a divide-by-5 counter (U_2). The counter is connected to the control elements of the analog multiplexer (U_1). The multiplexer can sample up to eight inputs and then provide a single composite output. As the divide-by-5 counter ripples through the analog switches, each analog input is sampled for a short time, and this sampled output is fed to an IRED driver. The IRED driver is, therefore, modulated by the composite signal passing through the analog multiplexer. The oscillator, operating at a frequency of 400 kHz, has a time period of 2.5 μs. This clock, as previously noted, indirectly controls the logic inputs of the multiplexer chip. At the end of the fifth count, the sampling is repeated.

Transistor Q_1 provides a 100- to 400-ns inhibiting pulse at the end of each gate's on period. This blank period prevents channel cross-talk.

The multiplexer in the transmitter receives binary pulses and converts these pulses to a decimal count

133

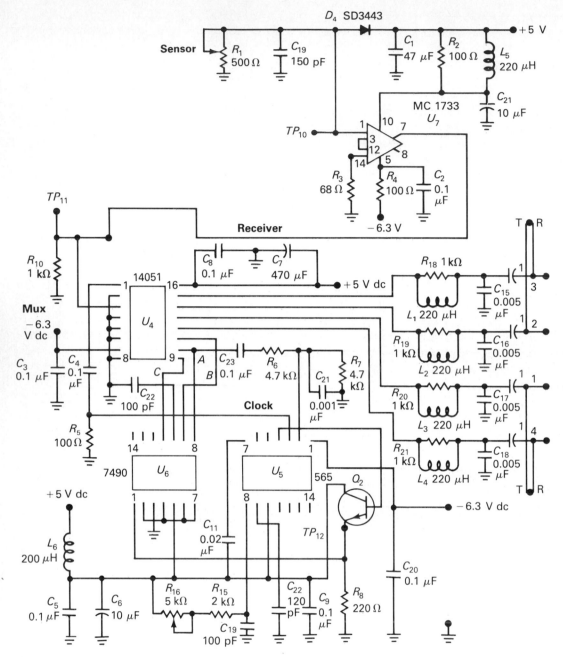

FIG. 29-5   Multiplexing receiver.

which controls channels 1 to 8. As noted, channel 5 was selected for synchronizing the system, and it contains the logic count for 5. When the count of 6 starts, the system recycles back to 1, while producing the triggering pulse which keeps the receiver synchronized with the transmitter.

The frequency of the transmitter clock (U_3) is determined by the 1.5-kΩ and 470-Ω resistors and the 47-pF capacitor. When the transmitter is multiplexing the various inputs, a composite signal is observed at the output (TP_9). The composite signal includes the synchronizing pulse and the four input channels. The

composite signal is used to drive the IRED of the transmitter whose output is coupled through a fiber-optic cable to the receiver.

For demonstration purposes, a function generator chip, LM566, has been provided. This chip and its *RC* filters provide four waveshapes which simulate information inputs, namely, a square wave, triangular wave, sine wave, and peak pulses. The four waveshapes can be selected via switches S_1 to S_4. If all switches remain open, the channels can be fed with other signal voltages. A stereo jack is shown in the circuit. Both channels of an FM radio can be multiplexed.

At the receiver, the output from the fiber-optic cable is received by the demultiplexing chip. The receiver's function is to sort out the various signals that have been transmitted and to direct these to the appropriate amplifiers. The receiver, whose circuit is shown in Fig. 29-5, functions in the reverse to the transmitter.

Whereas the transmitter receives numerous signals and combines them into a single composite signal, the receiver accepts the composite and divides it according to the proper channels.

The receiver has the following major sections:

1. An optical sensor and amplifier for converting the modulated light input to a composite signal voltage
2. A demultiplexer (U_4)
3. A logic sequencer (U_6)
4. A phase-lock loop comparator and filter (U_5)
5. A voltage-controlled oscillator (VCO) for transmitter synchronization (part of U_5).
6. Q_2, an emitter-follower (power driver) for U_6

The light signal received by the photodiode is amplified by a wideband IC chip (1733) so that it can activate the demultiplexer (U_4). The IC has a gain of 10 and a 120-MHz bandpass.

Once again, the logic control inputs of a multiplexer (marked A, B, and C) are driven by a 7490 (U_6). Again, a divide-by-5 function is used to control the multiplexer. The output of channel 5 and the B output of U_6 are fed to the phase-lock loop input (pins 2 and 3) of U_5. These two pulses operating at 80 kHz (depending on the original clock in the transmitter) are compared, and an error voltage is fed to the VCO portion of U_5 which, in turn, controls the divider chip U_6. By the start of the sixth count, the synchronizing pulse (from U_4) is received by the phase-lock loop and is compared to the output of the counter (U_6). An error voltage is developed, and it is this error voltage which is used to control the VCO (U_5) of the receiver. When the system is thus synchronized, or locked in, the positive-going pulse at the transmitter, which is on channel 5, will be used for synchronizing the receiver. The output of channel 1 will reflect the same signal which was fed into channel 1 of the transmitter, and likewise for channels 2, 3, and 4.

The clock adjust control R_{16} (5 kΩ) in the receiver shifts the oscillator frequency close enough for the error voltage to assume control (lock in).

In the above description, only five channels were utilized within the chip. However, the countdown or divide process could have been used to control numerous analog switches, and a composite of 50 to 100 signals could be multiplexed. Since there is a divide factor involved, the original clock frequency has to be increased as additional channels are added to the multiplexer. It is not unusual, therefore, to require that the clock frequency operate at several megahertz.

In this experiment an opportunity is provided for testing and evaluating the complete transmitter and receiver system. Several waveshapes from a generator are fed to the various channel inputs, and then the composite output from the receiver is observed and each channel tested separately.

The bandwidth of the signals that can be passed depends on the clock frequency as well as the bandwidth of the LED driver and the photodetector. The photosensor amplifier in this experiment has a bandpass of over 100 MHz; however, this is limited by the type of photodetector used and by the stray capacities that make up the circuitry.

In the evaluation of such a system, a variety of waveshapes can be observed. The composite wave, for example, is made up of the sync pulse plus the various signals passing through the four or more channels. A thorough familiarization is necessary in order to understand what the various waveshapes mean.

Since the synchronization of the receiver with the transmitter (in this design) depends on the amplitude of a positive-going pulse, it is important that none of the analog channels produce signals of equal amplitude to that of the synchronizing pulse. Should high-amplitude signals be present, the synchronizing pulse will be mixed with the multiplexed signals, and a loss of synchronization will take place. When the system is properly synchronized, it will automatically lock the receiver to the transmitter for the proper channel as soon as the system is energized.

The multiplexing system is used wherever it is desired to mix several signals on a single carrier. This technique is used in telephone communications, satellite communications, and telemetry of information. Obviously, the technique can also be used in instrumentation and the transfer of data from various transducers in an industrial process. The complex signal can be transmitted by wireless, a microwave link, or (as in this experiment) a fiber-optics cable. The latter is also immune to electromagnetic interference. The overall system consists of multiplexing, with the use of integrated-circuit chips, whose controllers are fed by logic dividers which trigger on the respective channels. The transmitter is controlled by a clock frequency whose rate is transmitted to the receiver, and the receiver utilizes a phase-lock loop (PLL) comparator to synchronize the clock of the receiver with that of the transmitter.

In voice communication, multiplexing is combined with the conversion of voice (analog) signals to digital pulses. The digital format usually takes the form of pulse-code modulation (PCM) in which the analog signal is quantized into a serial of eight pulses, and these pulses are multiplexed and transmitted over a single carrier. Single IC chips called *CODECs* are used for providing PCM and multiplexing. A full description of these chips is beyond the level of this experiment. Here only a multiplexing system is evaluated.

TESTS AND MEASUREMENTS

MATERIALS REQUIRED FOR EXPERIMENT

Active Devices
SE4352 IRED
SD3443 *IR* detector
MC14051 (2) 1-to-8 Decoder/encoder
7490 (2) Decade counter
NE555 Timer IC
NE566 VCO
NE565 PLL
MC1733 Op amp
2N3904 (2) NPN transistor
2N3905 PNP transistor
2N222 NPN transistor (driver)
2N4918 PNP transistor (power)
1N4148 SI diode
1N4736 6.3-V Zener diode

Resistors	Capacitors
10 Ω	100 pF
27 Ω	120 pF
47 Ω	150 pF
68 Ω	270 pF
100 Ω (3)	0.001 μF (2)
220 Ω	0.002 μF
470 Ω (4)	0.005 μF (4)
680 Ω	0.02 μF (2)
1 kΩ (9)	0.01 μF
1.5 kΩ (3)	0.1 μF (12)
2 kΩ (2)	1 μF (8)
2.2 kΩ (3)	10 μF (2)
3.9 kΩ	47 μF (2)
4.7 kΩ (2)	470 μF (2)
8.2 kΩ	1000 μF, 15 V
10.0 kΩ (3)	
22 kΩ	
33 kΩ	
1 MΩ (4)	
1-kΩ Potentiometer	
5-kΩ Potentiometer	

Other
Inductance, 220 μH (6)
Fiber cable, 50 cm, with connectors

In this evaluation, you will transmit multiplexed signals, receive, and observe them sorted according to their appropriate channels.

1. Arrange the circuits shown in Figs. 29-4 and 29-5. Both +5 and −6.3 V are required. See Fig. 29-4*b* for the −6.3-V supply. The IRED current is fixed at approximately 30 mA.
2. Use one or more of the generator signal sources. If external signal sources are used, the peak amplitudes should be under 2 V p-p. It is suggested that only one waveshape be used initially. Once it is observed, another can be added.
3. Turn on the power, and observe and sketch the input signal, in peak-to-peak volts, of input channels 1 to 4.
4. Check the frequency and amplitude of the transmitter clock at test point TP_8. What are the clock frequency at TP_8 and the amplitude?
5. Record the waveshapes at TP_5, TP_6, and TP_7. Set the time base of the oscilloscope to show at least four pulses at TP_5.
6. With only the sine wave present (switch 1a closed), record the composite signal at the transmitter's multiplexer output TP_9.
7. What is the frequency, in kilohertz, of the synchronizing pulse at TP_9?
8. Compare the amplitude of the signal V_{in} entering the multiplexer at channel 3 with its pulse amplitude V_o at TP_9. Are they approximately the same? Is there a voltage drop in the multiplexing switch? If so, how much?
9. Record the pattern across the 10-Ω resistor (TP_4) of the IRED driver TP_4.
10. How much peak current I_f flows through the IRED during synchronizing pulses?
11. How much pulse amplitude is present at the output of the sensing amplifier (TP_{11})?
12. Record the phase-lock loop (PLL) frequency of the receiver which is present at TP_{12}. Note that the PLL frequency is different from the transmitter clock frequency in question 4. When you are measuring with a counter or oscilloscope, the internal cable capacitance will alter the frequency of an oscillator. Use a low-capacitance cable. What is the value of TP_{12}?

NOTE: In the receiver, the 500-Ω potentiometer at the photodiode is normally set to approximately 250 Ω. This is adjusted as required by the signal strength.

13. Sketch the waveform and record the amplitude of the four output channels. If the signals are unstable, this indicates that the phase-lock loop has not locked its voltage-controlled oscillator in on the incoming transmitter synchronizing pulse. Adjusting R_{16} should cause the PLL to pull in and hold to

the transmitter pulse. A solid signal at each output channel will indicate that the system is properly locked in. You may observe that once locked in, R_{16} may be adjusted away from its setting without losing synchronization. How far this deviation is allowed is called the *locking range*, and R_{16} should be centered in this range. What are the peak-to-peak voltages at channels 1 to 4? In the receiver, *LC* and *RC* filters are used to remove the sampling changes at the outputs of channels 1, 2, 3, and 4. A filter would be added to each additional channel used, and the filter components would be computed for the switching frequency used.

14. Is the output waveshape similar to the input signal? Compare the input and output voltages for the four channels. Make a chart of V_{in} and V_o.
15. Determine the bandwidth of channel 1.
16. Could a voice signal be used on channel 1?
17. Describe briefly what the phase-lock loop does. In question 12 it was noted that the PLL frequency did not match the transmitter clock frequency because of the counter's internal capacitance loading and shifting of the oscillator clock frequency. Measure the frequency of the sync pulse which is derived by dividing the clock frequency by 5. Multiply this times 5, and compare it with the PLL frequency at TP_{12}. The readings must be concurrent since the transmitter clock frequency can drift. The PLL frequency matches the transmitter clock frequency as long as it is within its holding range. What are the frequencies, in kilohertz, at TP_9 and TP_{12}?
18. What would be the effect of lowering the clock frequency at the transmitter?
19. What must be done to the clock frequency if more channels are to be added?
20. If the signal frequency is 5 kHz, what should the minimum clock frequency be?
21. Perhaps the most fascinating application of the multiplex system is its use as a multichannel audio link. Depending on availability, you may feed the input channel(s) to the transmitter with voice signals from an audio amplifier or music from a radio station as fed from the earphone jack. It is necessary to amplify the receiver output channels so that they may be heard on a loudspeaker. If no amplifier is available, the output can be heard on headphones. The higher the impedance of the phones, the louder the sound will be. While listening to the voice modulation, slowly remove one of the light pipe connectors from its socket. Describe what occurs as the pipe is removed.
22. Stereo music from an FM receiver and/or tape cassette can be transmitted. Two jacks are provided at the input to the transmitter. Four stereo

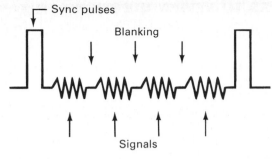

FIG. 29-6 Composite signal with four channels of transmission.

channels can be transmitted at the same time. Figure 29-6 shows how the waveshapes would appear at TP_9 and TP_{11}. A stereo amplifier is required at the output jacks of the receiver since the earphones present have too low an impedance.

23. The clock (U_3) provides a pulse to Q_1, which is an inverter. A positive pulse from Q_1 inhibits each channel at the end of its period. Measure and record the width of the inhibit pulses, in seconds, at the collector of Q_1.

REVIEW QUESTIONS

On a separate piece of paper, complete the following statements with the appropriate word or words.

1. The transmitter clock determines the period of _____.
2. The multiplexer channel has an on resistance of under _____ Ω.
3. To add more channels, the clock frequency must be _____.
4. To synchronize the receiver to the transmitter, a unique pulse is transmitted at the beginning or end of a(n) _____.
5. The sequencer controller provides a(n) _____ function.
6. In multiplexing, is the analog signal converted to digital pulses? _____.
7. If the signal output from the receiver channel is low in amplitude, what is suggested? _____.
8. If the clock of the transmitter is increased, what problem areas may arise? _____.
9. Could the multiplexer system be used with a microwave link? _____.
10. Does the MC17333 have a wide bandpass? _____.

ANSWERS TO REVIEW QUESTIONS

1. Each sample
2. 100
3. Increased
4. Frame
5. Countdown
6. No. The analog signal is sampled.
7. Add more gain to U_7 or to the desired channel. Increase the signal source to the transmitter. Reduce cable loss.
8. The bandpass of the link and/or receiver must be increased.
9. Yes
10. Yes

SIMPLEX AND DUPLEX SYSTEMS

SCOPE OF STUDY

In a simplex system, a single transmitter is linked by fiber to a single receiver. In a half-duplex system, a single fiber-optic cable provides for bidirectional flow; in a duplex system, two fiber cables are used for connecting each transmitter to its corresponding receiver. In this experiment, some basic principles used in simplex systems are explored.

OBJECTIVES

Upon completion of the assignments and evaluation of the data, you will be able to:

1. Describe the operation of a basic simplex data or communications link
2. Suggest applications for the use of simplex and duplex links
3. Test and evaluate a simplex system, given the instructional data
4. Communicate your understanding of the function of data links to other technical and nontechnical people
5. Research optoelectronic components for use in data links
6. Maintain and repair industrial/commercial equipment, given training on the overall operation of the equipment
7. Read and comprehend technical articles which describe the use of simplex and duplex links

BACKGROUND

Fiber-optic cables are used in short data links, audio-video interconnects, and in long lines which incorporate repeater stations. The repeater stations, which consist of a receiver and transmitter, are used to make up for cable losses, provide pulse shaping, and correct pulse overlapping. The system may be a one-way system (simplex), a duplex, or a half-duplex (the system can share the same cable). Regardless of the type of system used, emphasis must be given to the type of fiber cable used, its losses, and the matching of characteristics of the photodetector to those of the photoemitter. Flux budgeting, as described in Experiment 22, encompasses such overall planning. For short links (under 5 m)

low-cost plastic fiber cabling can be used with the losses being made up at the receiver by additional stage gain.

Although flux losses are an important factor, the noise on the line or the ability for the signal to be amplified above the noise is also important. Noise is generally classified into two types—white or pink noise. When noise in general is referred to, it is understood to be white noise. White noise is distributed rather equally across the selected bandwidth. Another type of noise is *impulse noise,* and this type of noise is a problem in pulse data transmission. White noise is a greater problem in analog transmission where the signal may not be much greater than the noise level.

Impulse noise, which is created by people, takes the form of sharp clicks or bursts of energy from various types of electric machines. Such impulses can cause errors in data transmission or improper synchronization. Although such electrical noise (EMI) does not affect the fiber cable, it can enter the system at either terminal. The fact that the fiber is immune to EMI is one of its many advantages. The signal-to-noise ratio S/N is expressed in terms of decibels, where

$$dB = 10 \log \frac{V_S}{V_N}$$

and $V =$ millivolts or microvolts.

The signal-to-noise ratio is generally expressed for a defined bandpass. In voice communications of 300 to 3400 Hz, the loss is 3 dB at the low end and 1.5 dB at the high end. The distance between repeaters will dictate the type of cable to be used. Other important factors are the number of signal channels, environmental conditions, and maintenance policy and procedures. Some systems may require modular units for ease in replacement. In general, the suitability of a fiber-optic system in an application will depend on the performance desired, cost of the fiber-optic cable, and its devices versus alternate types of systems. Quite often, short links or long lines needing low to medium bandwidths can be economically implemented. Wide-bandpass lines, used in data transmission, require special size fibers, usually of glass, and care must be exercised to minimize losses. At the time of this writing, both Bell Laboratories and British Telephone Laboratories have announced cable lengths, without repeaters, of up to 120 mi.

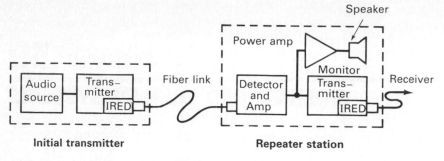

FIG. 30-1 Simplex repeater station.

TESTS AND MEASUREMENTS

MATERIALS REQUIRED FOR EXPERIMENT

Active Devices
SE4352 IRED, emitter with AMP sockets
SD3443 IRED, detector with AMP sockets
TL074 IC, op amp
2N3904 Transistors, NPN
2N4921 Transistors, power NPN
LM386 IC amplifier

Resistors	Capacitors
5 Ω	0.1 μF (3)
10 Ω (2)	10 μF (2)
560 Ω	47 μF (2)
3 kΩ	0.01 μF
10 kΩ (2)	0.05 μF
51 kΩ (2)	100 μF
100 kΩ	
220 kΩ	
1-kΩ Potentiometer	
10-kΩ Potentiometer	

Miscellaneous
50 cm, 1-mm fiber-optic cable with AMP plugs
Switches
Jacks
Small PM speaker

In this experiment, a simplex repeater system is demonstrated in simplified form. A radio receiver signal is used to modulate an optics transmitter and receiver for rebroadcasting.

1. Figure 30-2 shows the basic circuit for an optics transmitter and receiver. For the laboratory activity, the audio output from a transistor radio (earphone jack) is used as the modulating signal. The audio output from any AM or FM tuner can be used as the signal. If desired, digital pulses can also be used. The signal required is approximately 0.1 to 0.2 V p-p. The signal is fed into a light driver (modulator), as used in previous experiments.

   The infrared light output from the SE4352 emitter is connected to the receiver's photodiode via a 50-cm fiber-optic link. This link, in practice, could be over 1 km in length. The output from the repeater's amplifier (TL074) is fed to an audio power amplifier and speaker. The resistive load of the photodiode sensor and the gain of the photodiode's amplifier are dependent on the signal level. A gain of 10 to 15 may be adequate for this experiment.

   In a repeater station, the signal output from the detector's amplifier is used to modulate another transmitter whose output feeds on additional fiber-optic links. The amplifier could also be used to feed the station's power amplifier and speaker for local monitoring. With this background information, you are to make the necessary connections between the transmitter, receiver, a power amplifier, and speaker.

   The power supply voltages should be set as necessary.

2. The system should be functional when the power is turned on. Check the emitter current by measuring the voltage to ground at TP_4. This voltage should be 0.3 V (LED current equals 30 mA). Turn on the radio, select a station, and adjust the volume control.

   NOTE: Check the loudspeaker output to ensure that the amplifiers are not being overdriven.

3. Replace the radio with a 1-kHz sine wave set to

140

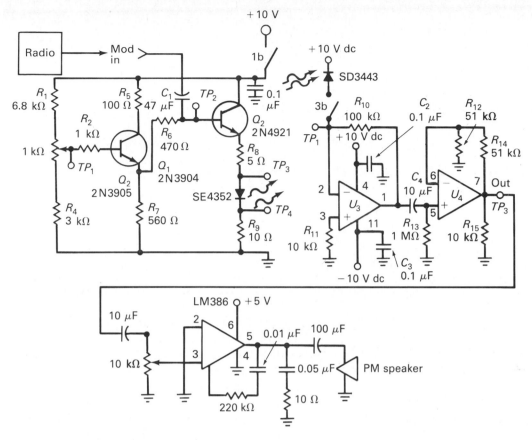

FIG. 30-2 Transmitter/receiver system.

0.1 V p-p. View the output voltage at TP_4 and the waveshape at TP_3. Determine what happens to the LED current when the modulation voltage is increased.

4. Determine the maximum driving voltage in the light driver that the amplifier (Q_2) can handle. Observe any flattening of the waveshape at TP_3.

5. Assume that the signal output of the receiver's amplifier is 1 V and the noise level is 2 mV. What is the signal-to-noise ratio S/N in decibels?

REVIEW QUESTIONS

On a separate piece of paper, complete the following statements with an appropriate word or words.

1. Generally _____-core fiber-optic cables have a lower loss than _____-core cables.
2. The signal voltage output from the receiver is determined by the _____ and the value of the _____.
3. Repeater stations are used to _____, _____, and _____.
4. Telephone repeaters are likely to be placed 2 mi apart whereas fiber-optic repeaters are currently being spaced _____ mi apart.

5. Do transistors and ICs produce noise? _____.
6. One way of reducing noise is by _____ the bandpass.
7. Does the use of printed circuits reduce noise? _____.
8. In a duplex system, can two people talk at the same time? _____.
9. A duplex system uses _____ fiber cable(s).
10. A half-duplex system uses _____ fiber cable(s) for two-way communications.

ANSWERS TO REVIEW QUESTIONS

1. Glass, plastic
2. Gain of the receiver, resistor load of the detector
3. Increase signal gain, enhance the signal to restore it to its original form, and improve S/N
4. 6 to 10
5. No
6. Reducing
7. No. Some circuits, through ground loops, can increase the noise; however, circuit noise is generally minimized.
8. Yes
9. Two
10. One

PART SEVEN

APPLICATIONS USING REFLECTIVE AND TRANSMISSIVE DEVICES

Reflective and transmissive devices are finding wide applications in a variety of instruments. One of the major applications of these devices is encountered in bar code reading, which is frequently seen at supermarket check-out counters. By reading the digital code placed on the consumer products, the price and the item are recorded and the inventory of the stock is adjusted.

The bar code consists of a series of narrow and wide lines which can be either read by a light wand or passed over a reflective surface which conceals a light emitter and a light sensor.

Bar code wands are also finding applications in learning equipment where they can be used with video disks and laboratory manuals to index immediately a position on the video disk for information recall.

Bar codes will also find industrial applications because they can automatically control the setting of machines and determine the process sequence to be followed in production.

In Part 7 two experiments are provided. The first deals with reflective and transmissive devices in general and the second with the bar code wand. The two experiments are predominately text-oriented, since the laboratory equipment would require sophisticated circuitry to make it fully operational.

REFLECTIVE AND TRANSMISSIVE SENSING

SCOPE OF STUDY

Optoelectronic devices find wide application in industrial, commercial, and consumer equipment where non-contact switching, sorting, or counting is required. In this study some typical applications are described. No experiment is included.

OBJECTIVES

Upon completion of the study and when presented with specific requirements, you will be able to:

1. Suggest circuit diagrams and optodevices for turning circuits off or on by using light-reflective or -transmissive techniques

2. Test and repair audio tape equipment which utilizes end-of-audio-tape switching

3. Describe how tachometers, for measuring shaft rotation, operate

BACKGROUND

The processing and displaying of digital information have numerous applications in the consumer, commercial, and industrial fields. The conversion of analog information into digital data can often be accomplished

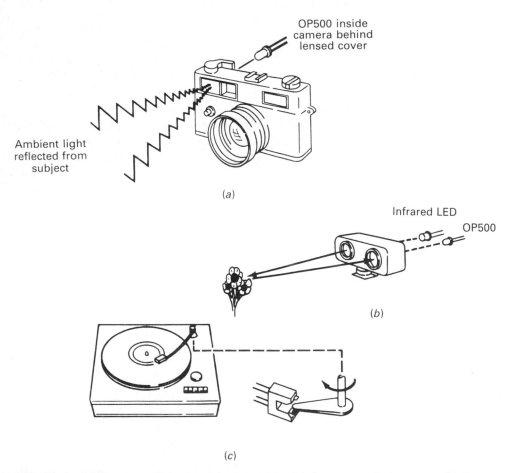

FIG. 31-1   Reflection and transmission sensing. (*a*) Automatic light meter. (*b*) Automatic focus control. (*c*) When the tone arm on a phonograph engages the oscillating grooves on the inside of the record, an opaque object connected to the underside of the tone arm interrupts the IR light path in the interrupter, triggering the return-to-rest or record change cycle.

by noncontact optical devices. The turning off of an audio cassette player, the reject at the end of a record, the location of pin holes in rolling sheet steel, the alignment of textile rolls during production, the shearing of paper, the register alignment of multicolor printing presses, and plastic or metal stamping are typical applications of light emitters and photosensors.

In transmissive switching, the light beam is broken or made to vary in amplitude as a result of the density of the material. In reflective applications, the light source is reflected off of a surface, and the amplitude and rate of pulsing are determined by the reflective pattern. The reflective light, for example, from a white-and-black bar pattern on a rotating shaft can be easily counted by digital circuitry. The noncontact tachometer is one such instrument which is used to determine the speed of rotation of a turning shaft, wheel, pulley, or cam.

In some applications, the directivity of the emitter and the responsivity of the detector are not critical. The presence or absence of light may be all that is required. In other applications where resolution is a critical requirement, special lenses may be required on the emitter, the photosensor, or both. Figure 31a, b, and c shows some applications of transmissive and reflective sensing.

In some applications, because of the distance between the emitter and sensor, special optical lenses may be required in order to focus the light beam on the distant photosensor. Figure 31-2 shows how a double convex lens is used to concentrate the light beam on a photosensor.

The placement and focal length of the lens are determined from the basic lens equation:

$$\frac{1}{f} = \frac{1}{d_s} + \frac{1}{d_r}$$

where d_s is the distance from source to the lens and d_R is the distance from the lens to the sensor. The symbol f is the focal length. The lens may be rated in terms of an f number or its numerical aperture (N.A.). The f number and N.A. rating of a lens are universally related. An $f/1.2$ lens has a larger N.A. value than a lens with an $f/2.0$ rating. The N.A. value of a lens approximates $D_L/2d_s$, where D_L is the diameter of the lens and d_s is the distance between the source and the lens. A lens with a rating of $f/1.7$ has a larger diameter than one

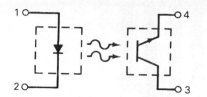

FIG. 31-3   Interrupter module.

rated at $f/2$. In some applications a lens is utilized at both the emitter and photosensor to obtain both a higher light intensity and more accurate focusing.

Monitoring the speed of rotation of a shaft or the interruption of a circuit by a tape, card, or other opaque material can be accomplished by a photocoupled transmissive interrupter module, as shown in Fig. 31-3. The sensor, on one side of the module, can be a photodiode, phototransistor, or photo-Darlington. When light is radiated on the sensor, the sensor is changed to the on state. Any interruption of light will shift the sensor to the off state. The module is low in cost, offers noncontact switching and fast switching speeds, is solid-state reliable, and is compatible with IC-controlled circuits.

A photosensitive module can also be made to operate in a reflective mode. In the reflective module, both the emitter and sensor are on the same side, and light is reflected back from white-and-black surfaces or bars. The white bar causes a light reflection while the black bar absorbs the light and thus reduces the reflected light. In bar code reading applications, the beam focusing must be less than the space between bars. Figure 31-4 shows a bar pattern which can be placed on a shaft or other rotating device for measuring the speed of rotation.

$$F = \frac{\text{Lines/rev}}{4\pi r}$$

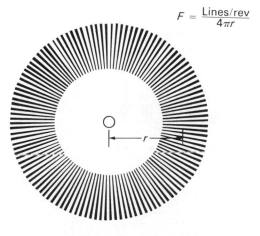

Spatial frequency

FIG. 31-4   Frequency code wheel. (*Courtesy of Hewlett-Packard Co.*)

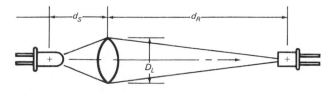

FIG. 31-2   Double-convex lens.

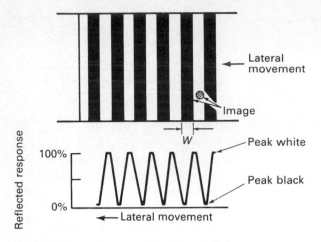

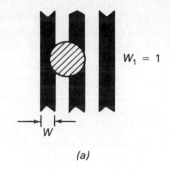

$W_1 = 1$

(a)

FIG. 31-5   Sensor output with reflectance.

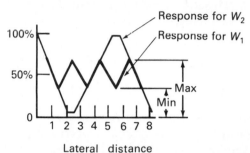

Lateral   distance

FIG. 31-6   Signal output with a large light-dot diameter.

The spatial frequency of the disk is determined by

$$F = \frac{\text{lines per revolution}}{4\pi r}$$

The radius r is taken from the center of the wheel to the midarc. The white bars generate n number of pulses per revolution, and this value, multiplied by 60, gives revolutions per minute. The frequency of the shaft rotation is determined from

$$f \text{ (Hz)} = \frac{n(\text{per minute})}{60}$$

where n is the number of line pairs (black and white).

Figure 31-5 shows the relationship between the sensor output and the white bars. The light image on the bars must be much smaller than the width of a bar, or the amplitude of the pulses will be reduced. In view of this, the light should be sharply focused on the wheel with a dot size equal to or less than 50 percent of the bar width. Figure 31-6 shows the effect on the output signal when the light dot bridges two bars. The sensor, as previously indicated, can be a photodiode, phototransistor, or Darlington device. The selection of the device

will depend on the frequency response, counting rate, or sensitivity required.

The amplifier which follows the sensor is a current-to-voltage converter. The sensor is back-biased, and the gain of the amplifier is determined by the feedback resistor R_f. Figure 31-7 shows a simplified circuit for the current-to-voltage converter. The pulsing output voltage V_o from the amplifier is determined from

$$V_o = I_s R_f$$

where I_s is the sensor current. Since the photodiode (or transistor) is changing in resistance, the gain of the amplifier is essentially R_f/R_i, where R_i is the instantaneous resistance value of the sensor. The amplifier, used as a current converter, is also referred to as a *transresistance amplifier*.

The total sensor current results from two light sources: the reflected light and the stray light, which

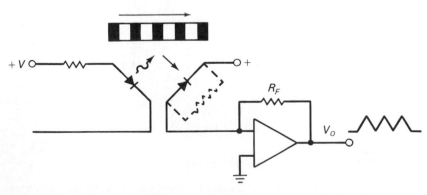

FIG. 31-7   Current-to-voltage conversion of the photodiode current.

bypasses the mechanical barriers to enter the sensor. The scatter of the LED light and even ambient light can enter into the sensing detector and change its operating characteristics.

The reflectiveness of the interrupter device depends upon:

1. The reflectiveness or transmissiveness of the material being monitored (that is, the contrast between reflection and absorption)
2. The size of the focused dot as compared to the reflective bars or the material's edge which is being monitored
3. Speed of the monitor
4. The ratio of the reflected light as compared to stray light
5. Optical, mechanical, and electrical effects

EXPERIMENT

No laboratory experiment is provided since the application involves mechanical alignment of moving parts. The use of the emitting diode, sensors, and amplifiers has been covered in previous experiments. The analog output from the sensor's amplifier can be coupled into a comparator for creating a pulse capable of interfacing with TTL circuitry. This type of circuitry has also been previously covered and so is not repeated.

SUMMARY

The study covered the use of emitter diodes and sensors, which were mechanically coupled so as to count rotations, serve as a noncontact switch, or monitor the movements of parts in production. The emitter and sensor can be individually located to monitor the operation, or they can be combined in a single unit called a photocoupled interrupter.

REVIEW QUESTIONS

Test your comprehension of the subject by answering the following questions. Answer True or False on a separate piece of paper.

1. If a separate light emitter and detector are used in a reflective application, the devices must be matched by the wavelength of operation.
2. The dot size should be very small compared to the bar size.
3. The smaller the dot size, as compared to the bar size, the larger will be the signal.
4. The light reflection from a white bar is greater than from a black bar.
5. The photosensor responds only to reflective light.
6. The numerical aperture of a focusing lens depends on its focal length and the diameter of the lens.
7. The f number of a lens relates to the ratio of the focal length to its diameter.
8. The smaller the f number, the smaller the lens.
9. The output voltage from a current-to-voltage converter depends on the ratio of R_f to R_i.
10. The output from a T-to-V converter can be converted to a square wave by a comparator.

ANSWERS TO REVIEW QUESTIONS

1. T
2. T
3. T
4. T
5. F
6. T
7. T
8. F
9. T
10. T

SCOPE OF STUDY

Bar code scanning is used for the labeling and pricing of products, for inventory control, and in production. Either the product is in motion, or a bar wand is passed over the reflective code. Both forms of motion produce a digital readout.

OBJECTIVES

Upon completion of the study, you will be familiar with the methods used in counting and the coding of materials. You should be able to:

1. Describe how optoelectronic devices are used in counting
2. Suggest basic circuitry that might prove useful in bar code reading
3. Troubleshoot basic bar code systems once you are familiar with the circuitry and components

BACKGROUND

The following material is a continuation of the topics covered in Experiment 31. In this section, the bar code and bar wand are discussed further.

In bar code scanning, digital data is stored in a black-and-white bar pattern. A standard, referred to as the *Universal Product Code* (UPC), establishes the bar-to-bar and space-to-space widths. The sensor which passes over the bar code provides an electrical output which is proportional to the width of the printed bar and its reflection. The bar-and-space width determines the width of the output pulse. The density of the reflected light determines the amplitude of the output pulses. Effective coding depends on the accuracy of encoding bars, spaces, and the size of the light dot. The smaller

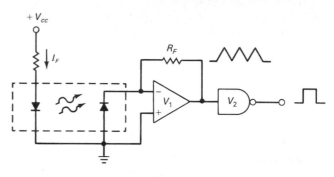

FIG. 32-2   Bar wand amplifier circuit.

the light-scanning image, the more rapid the pulse transition at the output. Once again, sharp change in the pulse amplitude is determined by sharp transitions between white and black bars.

Figure 32-1 shows a test pattern with bar widths of 0.3 mm. The optical sensor in a bar wand consists of a LED placed side by side with the sensor. A focusing lens is placed in front of each device. The sensor's output is amplified and digitalized for directly interfacing with logic circuitry. Figure 32-2 shows a simplified circuit for a bar wand amplifier.

In some bar wands the sensor, the amplifier, and the digitalizer are all included within the wand. The Hewlett Packard HEDS-3000, for example, is a complete wand which is capable of reading a resolution in bars of 0.3 mm (0.012 in). Figure 32-3 shows a sensor amplifier operated from a single voltage supply.

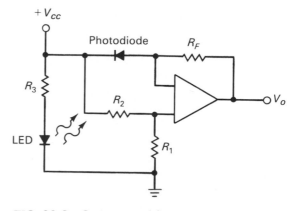

FIG. 32-3   Sensor amplifier circuit.

FIG. 32-1   Test bar encoded data.

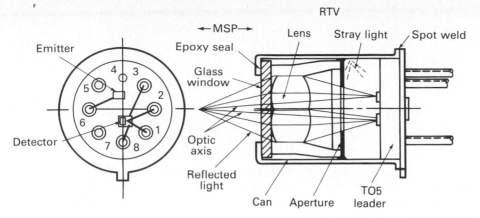

FIG. 32-4 Bar wand sensor. (MSP = maximum signal point.)

The output voltage V_o of the amplifier is determined from

$$V_o = \frac{V_{cc}}{1} + \frac{R_2}{R_1} - \frac{I_p}{R_f}$$

Resistors R_2 and R_1 provide linear biasing of the amplifier, and the output signal is a function of the sensor current (I_p) and the feedback resistor (R_f).

Figure 32-4 shows the construction of the HEDS-1000 optical sensor. Note that two carefully placed lenses are included within the header. When the light dot transverses the bars, the amplifier shown in Fig. 32-3 produces an output voltage which swings between 0.4 and 2.4 V. This voltage change can ensure the reliable operation of an analog comparator or logic gate.

Figure 32-5 shows a schematic for an amplifier system that will convert the bar and space widths of a bar code to TTL-compatible signals. The circuit uses the CA3130 as a transresistance amplifier for the HEDS-1000 photodiode. The output of the amplifier is applied to positive-peak (LM124-1) and negative-peak (LM124-2) detectors. Resistors R_1 and R_2 set the voltage reference (negative-going) input to the code comparator (LM124-3) at a voltage level which is halfway between the positive peak and the negative peak. The switching threshold is therefore at 50 percent of the peak-to-peak modulation. The noise gate (LM124-4) compares the negative peak to a voltage which is two diode voltage drops (D_1 and D_2) below the positive peak. Unless the peak-to-peak amplitude exceeds two diode drops, the G input of the 74LS75 remains low and Q cannot change. This ensures that the Q output of the 74LS75 will remain fixed unless the excursions at the output of the CA3130 are of an adequate amplitude (two diode drops) so that noise will not interfere.

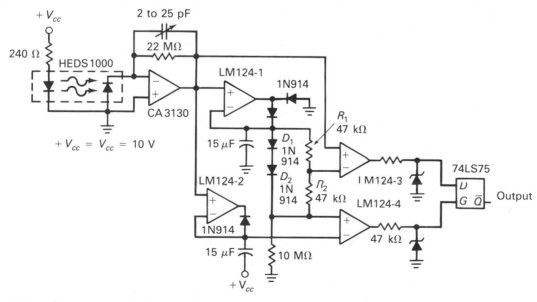

FIG. 32-5 Bar code scanner circuit. *(Courtesy of Hewlett-Packard Co.)*

Bar codes can contain such information as pricing, type of article, and quantity in stock or quantity to be deducted from the inventory. Much technical information exists on the various types of codes used, their layout, and the electronics needed to provide a direct readout. This study is limited to the concept of utilizing optodevices in converting bars and spaces to digital data.

SUMMARY

Bar scanning is both fast and accurate. A visit to a supermarket provides but one good example for the application of bar code scanning. Each article, moved quickly over the optically focused beam, is automatically recorded.

REVIEW QUESTIONS

On a separate piece of paper, complete the following statement with the appropriate word or words.

1. In bar code scanning, the light dot must be _____ than the black-and-white bars.
2. In scanning a bar code, the wand should be moved at a(n) _____ speed.
3. The bar wand may contain the _____, _____, _____, and _____.
4. The width of the bar (black and white) determines the _____ of the output pulse.
5. The amplitude of the output analog signal is determined by the _____ light.
6. The output analog voltage ranges from approximately _____ to _____ V.
7. In bar code reading, either the _____ or the _____ can move.
8. The printed bar code should be on material that produces a high order of _____.
9. The output circuit of the wand should be _____ compatible.
10. The gain of the current-voltage converter is related to _____ and _____.

ANSWERS TO REVIEW QUESTIONS

1. Smaller
2. Constant
3. Emitter, sensor, current voltage amplifier, and digitalizer
4. Width
5. Reflected
6. 0.4 to 2.4
7. Wand, bar code
8. Reflectance
9. TTL
10. I_p (diode current), R_f (feedback resistor)

MASTER LIST OF MATERIALS REQUIRED FOR EXPERIMENTS

MATERIALS REQUIRED

The components are standard commercial- and industrial-grade parts which are available at electronic parts outlets. The criteria for selection of these components are that they be readily available, of good quality, and of a size that could be easily handled by the experimenter. These components are given in the list that follows. The Science Instruments Company of Baltimore, Maryland, provides complete components kits, assembled panels, and a basic trainer.

COMPONENT TOLERANCES

Generally, resistors of 5 or 10 percent tolerance can be used. Capacitors may have 10 to 20 percent tolerance or more. In circuits where matched pairs of components are needed, you should check your component values and try to match them as closely as possible. Although 1 and 0.1 percent components are available, they are quite costly. Besides, it is good experience to learn when and how to find closely matching components from a low-tolerance batch since this is often necessary in fieldwork.

The parts required for each circuit are shown in the circuit diagrams. Resistors are all ¼ W unless otherwise indicated. Capacitors are rated at 25 working volts dc.

To perform the experiments, well-regulated power supplies are required. A well-regulated dual power supply with ± 15 V dc at 0.2 A and + 5 V at 0.2 A is required. The 15-V sources should be adjustable, and the ripple should be less than 5 mV with full load.

In addition, an ac supply of 2 to 5 V is required for several of the experiments.

In some cases, the use of alternate parts will result in answers that differ from the answers provided. The object in conducting the experiments is to give the student hands-on experience and a clearer understanding of the circuits. It is more important that the student learn how to obtain answers than to try to duplicate the exact results of the test.

Semiconductors

Quantity	Part number	Description	Alternate
1	SE4352	Optodiode	
2	SD3443	Phototransistor	ECG3031
2	VTA 3321/TIL414	Phototransistor	(RS)276-145 ECG3037
1	TIL111/4N37	Coupler, phototransistor	ECG3041
1	MOC3010	Coupler, triac driver	ECG3047
1	TIL119/4N30	Coupler, Darlington	ECG3084
1	LED visible	General-purpose indicator	ECG3022
1	LM386	Power amp IC	ECG823
1	TL081	BIFET op amp	ECG857M
2	TL074	Quad BIFET op amp	ECG859
1	MC1733	Differential op amp	ECG927D
1	NE555	Timer	ECG955M
1	NE556	Dual timer	ECG978
1	LM565	PLL	ECG989
1	LM566	VCO	ECG994
2	MC14051	8-Channel multiplexer	ECG4051B
1	SN7400	Quad NAND gate	ECG7400
1	SN7404	Hex inverter	ECG7404
2	SN7490	A system decade counter	ECG7490
1	MC75451	Peripheral driver AND	ECG75451B
1	MC75107	Dual-line receiver	DS75107
1	VT912	CDS photoresistor dark, $R = 20$ MΩ; illuminated, 2 fc, $R = 15$ kΩ	
1	MFOD300	Photo-Darlington	ECG3036
1	MFOD200	Phototransistor	ECG3031
1	MOC5010	Coupler, linear output	
1	SD4324	Opto Schmitt detector	ECG3039
8	IN4001	Diode, 1 A, 50 V	ECG116
3	IN4148	Diode, switching	ECG519
1	IN4736	Diode, zener, 6.8 V, 1 W	ECG5071A
1	IN4733	Diode, zener, 5 V, 1 W	ECG135A
1	2N2222	Transistor, NPN voltage amp	ECG123A
4	2N3904	Transistor, NPN voltage amp	ECG123AP
2	2N3905	Transistor, PNP voltage amp	ECG159
1	2N4918	Transistor, PNP power amp	ECG185
2	2N4921	Transistor, NPN power amp	ECG184
1	2N5062	SCR, 100 V, 0.8 A	ECG5402
1	2N4987	Silicon unilateral switch	ECG6404
1	ECG5655	Triac, 200 V, 0.8 A	2N6342

Resistors (All ¼ W Unless Stated)

Quantity	Part ID	Quantity	Part ID
1	5 Ω	1	18 kΩ
2	10 Ω	1	22 kΩ
2	27 Ω	1	27 kΩ
5	47 Ω	1	33 kΩ
1	68 Ω	5	47 kΩ
5	100 Ω	2	51 kΩ
1	120 Ω	3	100 kΩ
1	180 Ω	1	150 kΩ
1	220 Ω	1	220 kΩ
2	270 Ω	2	330 kΩ
3	330 Ω	1	470 kΩ
4	470 Ω	1	820 kΩ
1	510 Ω	4	1 MΩ
1	560 Ω	1	1.5 MΩ
1	680 Ω	1	2.2 MΩ
2	820 Ω	1	4.7 MΩ
9	1 kΩ	1	5.6 MΩ
1	1.2 kΩ	1	10 MΩ
3	1.5 kΩ	1	22 MΩ
2	2 kΩ		
3	2.2 kΩ	3	1-kΩ Potentiometer, linear
2	3.3 kΩ (3 kΩ)	1	5-kΩ Potentiometer, linear
1	3.6 kΩ	2	10-kΩ Potentiometer, linear
2	3.9 kΩ	1	100-kΩ Potentiometer, linear
2	4.7 kΩ	1	1-MΩ Potentiometer, linear
1	6.8 kΩ		
1	8.2 kΩ		
6	10 kΩ		
1	15 kΩ		

Capacitors

Quantity	Part ID	Description	Alternate
2	47 pF	Ceramic disk, 25 V, 20%	Mica, polystyrene
3	100 pF	Ceramic disk, 25 V, 20%	Mica, polystyrene
1	120 pF	Ceramic disk, 25 V, 20%	Mica, polystyrene
1	150 pF	Ceramic disk, 25 V, 20%	Mica, polystyrene
2	220 pF	Ceramic disk, 25 V, 20%	Mica, polystyrene
1	270 pF	Ceramic disk, 25 V, 20%	Mica, polystyrene
1	330 pF	Ceramic disk, 25 V, 20%	Mica, polystyrene
1	390 pF	Ceramic disk, 25 V, 20%	Mica, polystyrene
1	560 pF	Ceramic disk, 25 V, 20%	Mica, polystyrene
2	0.001 μF	Ceramic disk, 25 V, 20%	Mylar, polystyrene
2	0.002 μF	Ceramic disk, 25 V, 20%	Mylar, polystyrene
2	0.003 μF	Ceramic disk, 25 V, 20%	Mylar, polystyrene
4	0.005 μF	Ceramic disk, 25 V, 20%	Mylar, polystyrene
4	0.01 μF	Ceramic disk, 25 V, 20%	Mylar, polystyrene
2	0.02 μF	Ceramic disk, 25 V, 20%	Mylar, polystyrene
1	0.03 μF	Ceramic disk, 25 V, 20%	Mylar, polystyrene
1	0.05 μF	Ceramic disk, 25 V, 20%	Mylar, polystyrene
12	0.1 μF	Ceramic disk, 25 V, 20%	Mylar, electrolytic
2	0.22 μF	Ceramic disk, 25 V, 20%	Mylar, electrolytic
8	1.0 μF	Electrolytic, 25 V, 20%	Tantalum, tantalytic
2	10 μF	Electrolytic, 25 V, 20%	Tantalum, tantalytic
1	22 μF	Electrolytic, 25 V, 20%	Tantalum, tantalytic
2	47 μF	Electrolytic, 25 V, 20%	Tantalum, tantalytic
1	100 μF	Electrolytic, 25 V, 20%	Tantalum, tantalytic
2	470 μF	Electrolytic, 25 V	Tantalum
1	1000 μF	Electrolytic, 25 V	Tantalum

Inductors

Quantity	Part ID	Description
2	22 μH	Choke
6	220 μH	Choke
1	10 mH	Choke

Miscellaneous

Quantity	Part ID	Description	Alternate
3 m	Cable	Fiber optic, OE1040	Science Instruments Co. (SIC)
3 m	Cable	Fiber optic, 4001	Science Instruments Co.
50 cm	Cable	Fiber optic, OE1040	Science Instruments Co.
		Optoconnectors	
8	228041-1	Fiber-optic cable plug	Science Instruments Co.
8	228046-1	Retention clip	Science Instruments Co.
2	228040-1	Cable mount for TO-92 package*	Science Instruments Co.
		or	
2	228043-1	Cable mount for TO-18 package*	Science Instruments Co.
1	Relay	3 to 5 V, SPST	
2		Dip switches, 4PST	
1	Speaker	8 Ω, 1 W	
6 in	0.25-in-diameter	Heat shrink (black)	
		Dip sockets	
		Hook-up wire	

* Use ones that fit optodevices chosen.

Optional Parts List for Experiment 13

Quantity	Part ID	Description
1		100-W 120-V Lamp and socket
1		120/120 Isolation transformer
2	W005	Bridge rectifier's 200 PIV 1A
1	1N5457	Zener diode, 17 V, ½ W
1	ECG5437	SCR, 6 A, 200 V
1	ECG5608	Triac, 6 A, 200 V ac

MANUFACTURERS' DATA SHEETS

CONTENTS

Motorola Semiconductors 1N4001 through 1N4007   156
Motorola Semiconductors 1N4728,A through 1N4764,A   159
SD3443-1 through SD3443-3 LEDs, SE4352 LED   162
Motorola Semiconductors 2N3903, 2N3904   163
Motorola Semiconductors 2N3905, 2N3906   165
Motorola Semiconductors 2N4918 through 2N4920   168
Motorola Semiconductors 2N4921 through 2N4923   170
Motorola Semiconductors 4N29, 4N29A, 4N30, 4N31, 4N32, 4N32A, 4N33   172
Motorola Semiconductors 4N35 through 4N37   175
Motorola Semiconductors MC1733, MC1733C   178
Motorola Semiconductors MC14051B through MC14053B   181
SD4324 Schmitt Receiver   184
Motorola Semiconductors MC55107, MC55108, MC75107, MC75108   185
Motorola Semiconductors MFOD200   188
Motorola Semiconductors MFOD300   190
Motorola Semiconductors MOC3009 through MOC3012   192
Motorola Semiconductors MOC5010   195
National Semiconductor LM386   197
National Semiconductor LM555/LM555C   199
National Semiconductor LM556/LM556C   201
National Semiconductor LM565/LM565C   203
National Semiconductor LM566/LM566C   205
SIECOR 144 Fat Fiber Cable   207
EOTec Corporation Plastic Clad Silica Optical Fiber   208
Signetics Gates 54/7400, LS00, S00   209
Signetics Inverters 54/7404, LS04, S04   211
Signetics Counters 54/7490, LS90   213

MOTOROLA Semiconductors
BOX 20912 • PHOENIX, ARIZONA 85036

Designers Data Sheet

1N4001 thru 1N4007

LEAD MOUNTED SILICON RECTIFIERS

50-1000 VOLTS
DIFFUSED JUNCTION

"SURMETIC"▲ RECTIFIERS

. . . subminiature size, axial lead mounted rectifiers for general-purpose low-power applications.

Designers Data for "Worst Case" Conditions

The Designers▲ Data Sheets permit the design of most circuits entirely from the information presented. Limit curves — representing boundaries on device characteristics — are given to facilitate "worst case" design.

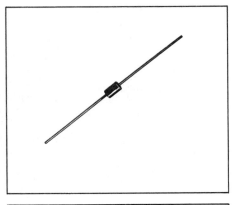

*MAXIMUM RATINGS

Rating	Symbol	1N4001	1N4002	1N4003	1N4004	1N4005	1N4006	1N4007	Unit
Peak Repetitive Reverse Voltage Working Peak Reverse Voltage DC Blocking Voltage	V_{RRM} V_{RWM} V_R	50	100	200	400	600	800	1000	Volts
Non-Repetitive Peak Reverse Voltage (halfwave, single phase, 60 Hz)	V_{RSM}	60	120	240	480	720	1000	1200	Volts
RMS Reverse Voltage	$V_{R(RMS)}$	35	70	140	280	420	560	700	Volts
Average Rectified Forward Current (single phase, resistive load, 60 Hz, see Figure 8, $T_A = 75^oC$)	I_O	1.0							Amp
Non-Repetitive Peak Surge Current (surge applied at rated load conditions, see Figure 2)	I_{FSM}	30 (for 1 cycle)							Amp
Operating and Storage Junction Temperature Range	T_J, T_{stg}	−65 to +175							^oC

*ELECTRICAL CHARACTERISTICS

Characteristic and Conditions	Symbol	Typ	Max	Unit
Maximum Instantaneous Forward Voltage Drop ($i_F = 1.0$ Amp, $T_J = 25^oC$) Figure 1	v_F	0.93	1.1	Volts
Maximum Full-Cycle Average Forward Voltage Drop ($I_O = 1.0$ Amp, $T_L = 75^oC$, 1 inch leads)	$V_{F(AV)}$	–	0.8	Volts
Maximum Reverse Current (rated dc voltage) $T_J = 25^oC$ $T_J = 100^oC$	I_R	0.05 1.0	10 50	μA
Maximum Full-Cycle Average Reverse Current ($I_O = 1.0$ Amp, $T_L = 75^oC$, 1 inch leads)	$I_{R(AV)}$	–	30	μA

*Indicates JEDEC Registered Data.

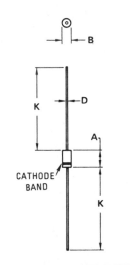

CATHODE BAND

DIM	MILLIMETERS		INCHES	
	MIN	MAX	MIN	MAX
A	5.97	6.60	0.235	0.260
B	2.79	3.05	0.110	0.120
D	0.76	0.86	0.030	0.034
K	27.94	–	1.100	–

CASE 59-04
Does Not Conform to DO-41 Outline.

MECHANICAL CHARACTERISTICS

CASE: Void free, Transfer Molded
MAXIMUM LEAD TEMPERATURE FOR SOLDERING PURPOSES: 350^oC, 3/8" from case for 10 seconds at 5 lbs. tension
FINISH: All external surfaces are corrosion-resistant, leads are readily solderable
POLARITY: Cathode indicated by color band
WEIGHT: 0.40 Grams (approximately)

▲Trademark of Motorola Inc.

© MOTOROLA INC., 1975

DS 6015 R3

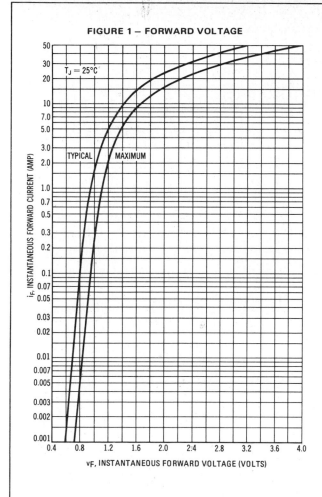

FIGURE 1 — FORWARD VOLTAGE

v_F, INSTANTANEOUS FORWARD VOLTAGE (VOLTS)

i_F, INSTANTANEOUS FORWARD CURRENT (AMP)

$T_J = 25°C$

TYPICAL   MAXIMUM

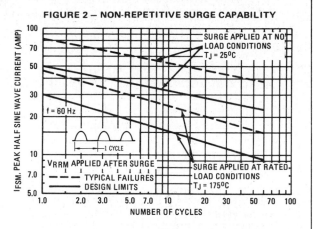

FIGURE 2 — NON-REPETITIVE SURGE CAPABILITY

I_{FSM}, PEAK HALF SINE WAVE CURRENT (AMP)

NUMBER OF CYCLES

SURGE APPLIED AT NO LOAD CONDITIONS $T_J = 25°C$

$f = 60$ Hz

1 CYCLE

V_{RRM} APPLIED AFTER SURGE

--- TYPICAL FAILURES
— DESIGN LIMITS

SURGE APPLIED AT RATED LOAD CONDITIONS $T_J = 175°C$

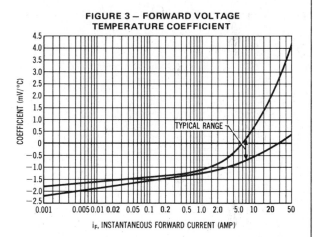

FIGURE 3 — FORWARD VOLTAGE TEMPERATURE COEFFICIENT

COEFFICIENT (mV/°C)

i_F, INSTANTANEOUS FORWARD CURRENT (AMP)

TYPICAL RANGE

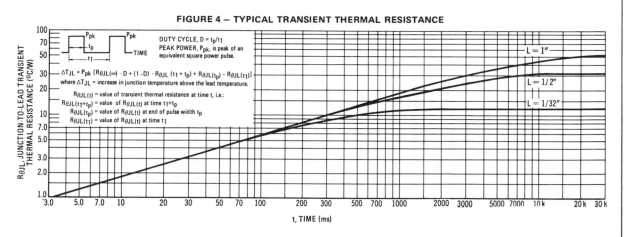

FIGURE 4 — TYPICAL TRANSIENT THERMAL RESISTANCE

$R_{\theta JL}$, JUNCTION-TO-LEAD TRANSIENT THERMAL RESISTANCE (°C/W)

t, TIME (ms)

$L = 1''$

$L = 1/2''$

$L = 1/32''$

DUTY CYCLE, $D = t_p/t_1$
PEAK POWER, P_{pk}, is peak of an equivalent square power pulse.

P_{pk} P_{pk}
t_p TIME
t_1

$\Delta T_{JL} = P_{pk} [R_{\theta JL}(\infty) \cdot D + (1 - D) \cdot R_{\theta JL}(t_1 + t_p) + R_{\theta JL}(t_p) - R_{\theta JL}(t_1)]$

where ΔT_{JL} = increase in junction temperature above the lead temperature.

$R_{\theta JL}(t)$ = value of transient thermal resistance at time t, i.e.:
$R_{\theta JL}(t_1 + t_p)$ = value of $R_{\theta JL}(t)$ at time $t_1 + t_p$
$R_{\theta JL}(t_p)$ = value of $R_{\theta JL}(t)$ at end of pulse width t_p
$R_{\theta JL}(t_1)$ = value of $R_{\theta JL}(t)$ at time t_1

The temperature of the lead should be measured using a thermocouple placed on the lead as close as possible to the tie point. The thermal mass connected to the tie point is normally large enough so that it will not significantly respond to heat surges generated in the diode as a result of pulsed operation once steady-state conditions are achieved. Using the measured value of T_L, the junction temperature may be determined by:

$$T_J = T_L + \Delta T_{JL}.$$

 MOTOROLA *Semiconductor Products Inc.*

CURRENT DERATING DATA

FIGURE 5 — FORWARD POWER DISSIPATION

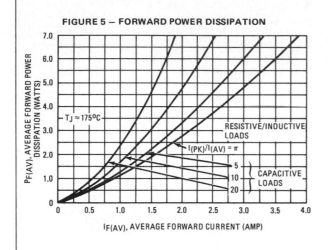

FIGURE 6 — EFFECT OF LEAD LENGTHS, RESISTIVE LOAD

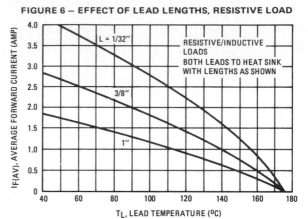

FIGURE 7 — 3/8″ LEAD LENGTH, VARIOUS LOADS

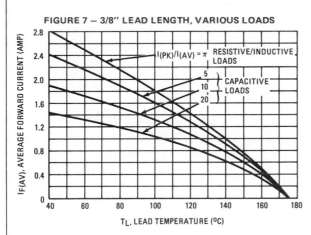

FIGURE 8 — PRINTED CIRCUIT BOARD MOUNTING — VARIOUS LOADS

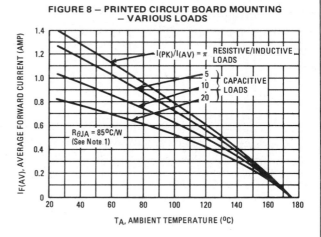

FIGURE 9 — STEADY-STATE THERMAL RESISTANCE

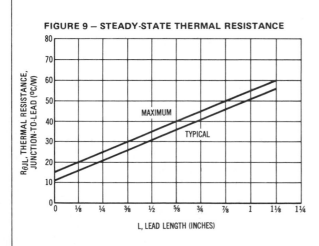

NOTE 1

Data shown for thermal resistance junction-to-ambient ($R_{\theta JA}$) for the mountings shown is to be used as typical guideline values for preliminary engineering or in case the tie point temperature cannot be measured

TYPICAL VALUES FOR $R_{\theta JA}$ IN STILL AIR

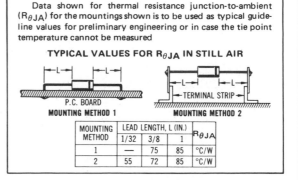

MOUNTING METHOD	LEAD LENGTH, L (IN.)			$R_{\theta JA}$
	1/32	3/8	1	
1	—	75	85	°C/W
2	55	72	85	°C/W

 MOTOROLA *Semiconductor Products Inc.*

MOTOROLA

Semiconductors

BOX 20912 • PHOENIX, ARIZONA 85036

Designers▲ Data Sheet

ONE WATT HERMETICALLY SEALED GLASS SILICON ZENER DIODES

- Complete Voltage Range — 2.4 to 100 Volts
- DO-41 Package — Smaller than Conventional DO-7 Package
- Double Slug Type Construction
- Metallurgically Bonded Construction
- Nitride Passivated Die

Designer's Data for "Worst Case" Conditions

The Designers▲ Data sheets permit the design of most circuits entirely from the information presented. Limit curves — representing boundaries on device characteristics — are given to facilitate "worst case" design.

1N4728, A
thru
1N4764, A

1.0 WATT

ZENER REGULATOR DIODES

3.3–100 VOLTS

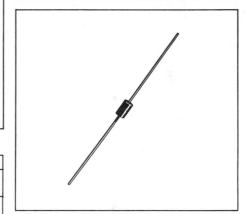

---

*MAXIMUM RATINGS

Rating	Symbol	Value	Unit
DC Power Dissipation @ T_A = 50°C	P_D	1.0	Watt
Derate above 50°C		6.67	mW/°C
Operating and Storage Junction Temperature Range	T_J, T_{stg}	–65 to +200	°C

MECHANICAL CHARACTERISTICS

CASE: Double slug type, hermetically sealed glass

MAXIMUM LEAD TEMPERATURE FOR SOLDERING PURPOSES: 230°C, 1/16" from case for 10 seconds

FINISH: All external surfaces are corrosion resistant with readily solderable leads.

POLARITY: Cathode indicated by color band. When operated in zener mode, cathode will be positive with respect to anode.

MOUNTING POSITION: Any

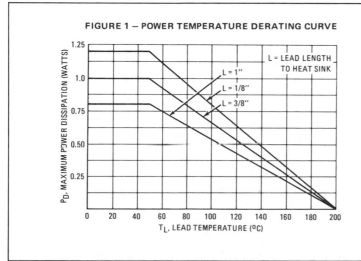

FIGURE 1 — POWER TEMPERATURE DERATING CURVE

L = LEAD LENGTH TO HEAT SINK
L = 1"
L = 1/8"
L = 3/8"

P_D, MAXIMUM POWER DISSIPATION (WATTS)

T_L, LEAD TEMPERATURE (°C)

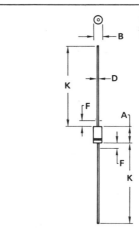

NOTE:
1. POLARITY DENOTED BY CATHODE BAND
2. LEAD DIAMETER NOT CONTROLLED WITHIN "F" DIMENSION.

DIM	MILLIMETERS		INCHES	
	MIN	MAX	MIN	MAX
A	4.07	5.20	0.160	0.205
B	2.04	2.71	0.080	0.107
D	0.71	0.86	0.028	0.034
F	—	1.27	—	0.050
K	27.94	—	1.100	—

All JEDEC dimensions and notes apply.

CASE 59-03
(DO-41)

---

*Indicates JEDEC Registered Data
▲Trademark of Motorola Inc.

©MOTOROLA INC., 1978

DS 7039 R1

JEDEC Type No. (Note 1)	Nominal Zener Voltage V_Z @ I_{ZT} Volts (Notes 2 and 3)	Test Current I_{ZT} mA	Maximum Zener Impedance (Note 4)			Leakage Current		Surge Current @ T_A = 25°C i_r – mA (Note 5)
			Z_{ZT} @ I_{ZT} Ohms	Z_{ZK} @ I_{ZK} Ohms	I_{ZK} mA	I_R μA Max	V_R Volts	
1N4728	3.3	76	10	400	1.0	100	1.0	1380
1N4729	3.6	69	10	400	1.0	100	1.0	1260
1N4730	3.9	64	9.0	400	1.0	50	1.0	1190
1N4731	4.3	58	9.0	400	1.0	10	1.0	1070
1N4732	4.7	53	8.0	500	1.0	10	1.0	970
1N4733	5.1	49	7.0	550	1.0	10	1.0	890
1N4734	5.6	45	5.0	600	1.0	10	2.0	810
1N4735	6.2	41	2.0	700	1.0	10	3.0	730
1N4736	6.8	37	3.5	700	1.0	10	4.0	660
1N4737	7.5	34	4.0	700	0.5	10	5.0	605
1N4738	8.2	31	4.5	700	0.5	10	6.0	550
1N4739	9.1	28	5.0	700	0.5	10	7.0	500
1N4740	10	25	7.0	700	0.25	10	7.6	454
1N4741	11	23	8.0	700	0.25	5.0	8.4	414
1N4742	12	21	9.0	700	0.25	5.0	9.1	380
1N4743	13	19	10	700	0.25	5.0	9.9	344
1N4744	15	17	14	700	0.25	5.0	11.4	304
1N4745	16	15.5	16	700	0.25	5.0	12.2	285
1N4746	18	14	20	750	0.25	5.0	13.7	250
1N4747	20	12.5	22	750	0.25	5.0	15.2	225
1N4748	22	11.5	23	750	0.25	5.0	16.7	205
1N4749	24	10.5	25	750	0.25	5.0	18.2	190
1N4750	27	9.5	35	750	0.25	5.0	20.6	170
1N4751	30	8.5	40	1000	0.25	5.0	22.8	150
1N4752	33	7.5	45	1000	0.25	5.0	25.1	135
1N4753	36	7.0	50	1000	0.25	5.0	27.4	125
1N4754	39	6.5	60	1000	0.25	5.0	29.7	115
1N4755	43	6.0	70	1500	0.25	5.0	32.7	110
1N4756	47	5.5	80	1500	0.25	5.0	35.8	95
1N4757	51	5.0	95	1500	0.25	5.0	38.8	90
1N4758	56	4.5	110	2000	0.25	5.0	42.6	80
1N4759	62	4.0	125	2000	0.25	5.0	47.1	70
1N4760	68	3.7	150	2000	0.25	5.0	51.7	65
1N4761	75	3.3	175	2000	0.25	5.0	56.0	60
1N4762	82	3.0	200	3000	0.25	5.0	62.2	55
1N4763	91	2.8	250	3000	0.25	5.0	69.2	50
1N4764	100	2.5	350	3000	0.25	5.0	76.0	45

*Indicates JEDEC Registered Data.

NOTE 1 — Tolerance and Type Number Designation. The JEDEC type numbers listed have a standard tolerance on the nominal zener voltage of ±10%. A standard tolerance of ±5% on individual units is also available and is indicated by suffixing "A" to the standard type number.

NOTE 2 — Specials Available Include:

A. Nominal zener voltages between the voltages shown and tighter voltage tolerances,
B. Matched sets.

For detailed information on price, availability, and delivery, contact your nearest Motorola representative.

NOTE 3 — Zener Voltage (V_Z) Measurement. Motorola guarantees the zener voltage when measured at 90 seconds while maintaining the lead temperature (T_L) at 30°C ± 1°C, 3/8" from the diode body.

NOTE 4 — Zener Impedance (Z_Z) Derivation. The zener impedance is derived from the 60 cycle ac voltage, which results when an ac current having an rms value equal to 10% of the dc zener current (I_{ZT} or I_{ZK}) is superimposed on I_{ZT} or I_{ZK}.

NOTE 5 — Surge Current (i_r) Non-Repetitive. The rating listed in the electrical characteristics table is maximum peak, non-repetitive, reverse surge current of 1/2 square wave or equivalent sine wave pulse of 1/120 second duration superimposed on the test current, I_{ZT}, per JEDEC registration; however, actual device capability is as described in Figures 4 and 5.

APPLICATION NOTE

Since the actual voltage available from a given zener diode is temperature dependent, it is necessary to determine junction temperature under any set of operating conditions in order to calculate its value. The following procedure is recommended: Lead Temperature, T_L, should be determined from

$$T_L = \theta_{LA} P_D + T_A$$

θ_{LA} is the lead-to-ambient thermal resistance (°C/W) and P_D is the power dissipation. The value for θ_{LA} will vary and depends on the device mounting method. θ_{LA} is generally 30 to 40°C/W for the various clips and tie points in common use and for printed circuit board wiring.

The temperature of the lead can also be measured using a thermocouple placed on the lead as close as possible to the tie point. The thermal mass connected to the tie point is normally large enough so that it will not significantly respond to heat surges generated in the diode as a result of pulsed operation once steady-state conditions are achieved. Using the measured value of T_L, the junction temperature may be determined by:

$$T_J = T_L + \Delta T_{JL}.$$

ΔT_{JL} is the increase in junction temperature above the lead temperature and may be found as follows:

$$\Delta T_{JL} = \theta_{JL} P_D$$

θ_{JL} may be determined from Figure 3 for dc power conditions. For worst-case design, using expected limits of I_Z, limits of P_D and the extremes of $T_J (\Delta T_J)$ may be estimated. Changes in voltage, V_Z, can then be found from:

$$\Delta V = \theta_{VZ} \Delta T_J$$

θ_{VZ}, the zener voltage temperature coefficient, is found from Figure 2.

Under high power-pulse operation, the zener voltage will vary with time and may also be affected significantly by the zener resistance. For best regulation, keep current excursions as low as possible.

Surge limitations are given in Figure 5. They are lower than would be expected by considering only junction temperature, as current crowding effects cause temperatures to be extremely high in small spots resulting in device degradation should the limits of Figure 5 be exceeded.

 MOTOROLA *Semiconductor Products Inc.*

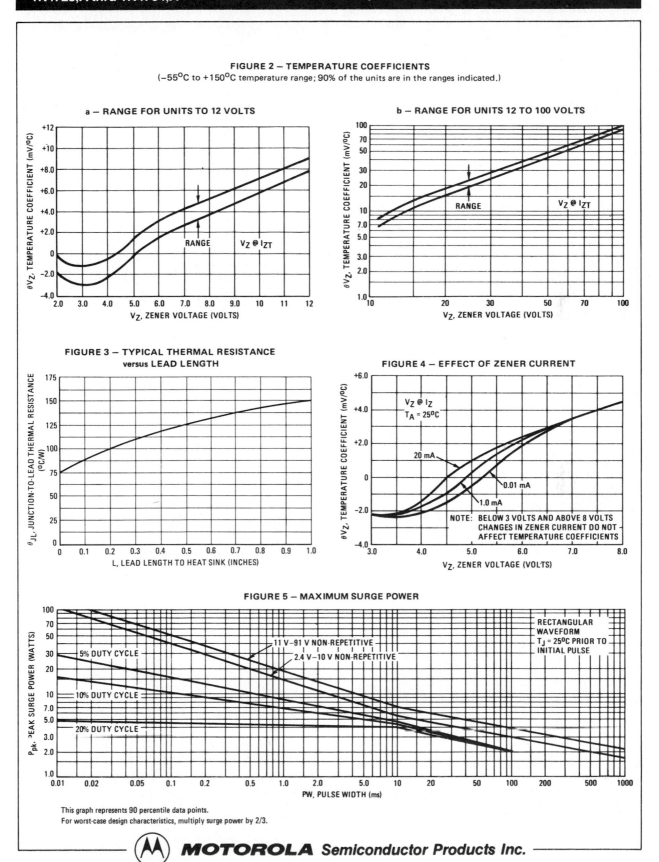

FIGURE 2 – TEMPERATURE COEFFICIENTS
(−55°C to +150°C temperature range; 90% of the units are in the ranges indicated.)

a – RANGE FOR UNITS TO 12 VOLTS

b – RANGE FOR UNITS 12 TO 100 VOLTS

FIGURE 3 – TYPICAL THERMAL RESISTANCE versus LEAD LENGTH

FIGURE 4 – EFFECT OF ZENER CURRENT

NOTE: BELOW 3 VOLTS AND ABOVE 8 VOLTS CHANGES IN ZENER CURRENT DO NOT AFFECT TEMPERATURE COEFFICIENTS

FIGURE 5 – MAXIMUM SURGE POWER

This graph represents 90 percentile data points.
For worst-case design characteristics, multiply surge power by 2/3.

MOTOROLA *Semiconductor Products Inc.*

SD3443-1 through SD3443-3 LEDs

Characteristics	Package	Light current			Dark current		Collector breakdown	Emitter breakdown	Saturation voltage	Light current rise time	Angular response
Test condition		$V_{CE} = 5$ V, H as shown			$H = 0$, V_{CE} as shown		$I_C = 100\ \mu A$	$I_C = 100\ \mu A$	$I_C = 0.4$ mA	$R_L = 1000\ \Omega$, $V_{CC} = 5$ V, $I_L = 1$ mA	3
Symbol		I_L			I_D		BV_{CEO}	BV_{ECO}	$V_{CE(sat)}$	t_r	$\emptyset$
Units		mA			nA		V	V	V	μs	degrees
		min.	max.	H	max.	V_{CE}	min.	min.	typ.	typ.	typ.
SD3443-1	Hermetic	0.5	5	100	10	30	7	0.2	6(b)	90	
SD3443-2	Hermetic	1.0	5	100	10	30	7	0.2	6(b)	90	
SD3443-3	Hermetic	2.0	5	100	10	30	7	0.2	8(b)	90	

SE 4352 LED

ELECTRO-OPTICAL CHARACTERISTIC ($T_C = 25\,^{\circ}$C UNLESS OTHERWISE SPECIFIED)

Parameter	Test Condition	Symbol	Min	Typ	Max	Units
Forward Drop	$I_f = 50$mA	V_f		1.6	2.0	volts
Series Resistance		R_s		1.6		Ω
Device Capacitance	$V_R = 1$V	C_T		800		p_f
Power Output	$I_f = 50$mA Aperture = 1mm NA = .5					
−002			175	400		μW
−003			350	700		μW
Response Time	1VDC Bias, I PEAK = 20mA	t_f		12	20	ns
Peak Output Wavelength	$I_f = 50$mA	λp		820		nm
Spectral Bandwidth	$I_f = 50$mA	Δd		35		nm
V_f Temperature Coefficient		$\Delta V_f / \Delta T$		−1.7		mV/$^{\circ}$C
P_o Temperaure Coefficient	$I_f = 50$mA			−.012		dB/$^{\circ}$C
λ Temperature Coefficient		$\Delta \lambda / \Delta T$		.35		nm/$^{\circ}$C
Thermal Resistance		θ		500		$^{\circ}$C/W
Operating Temperature			−40		100	$^{\circ}$C

2N**3903** (SILICON)
2N**3904**

NPN SILICON
SWITCHING & AMPLIFIER
TRANSISTORS

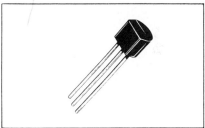

NPN SILICON ANNULAR TRANSISTORS

. . . designed for general purpose switching and amplifier applications and for complementary circuitry with types 2N3905 and 2N3906.

- Collector-Emitter Breakdown Voltage — BV_{CEO} = 40 Vdc (Min)
- Current Gain Specified from 100 μA to 100 mA
- Complete Switching and Amplifier Specifications
- Low Capacitance — C_{ob} = 4.0 pF (Max)

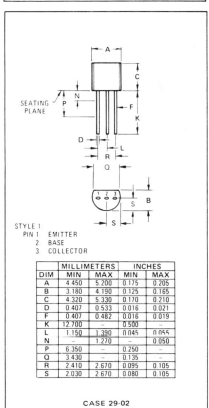

*MAXIMUM RATINGS

Rating	Symbol	Value	Unit
Collector-Base Voltage	V_{CB}	60	Vdc
Collector-Emitter Voltage	V_{CEO}	40	Vdc
Emitter-Base Voltage	V_{EB}	6.0	Vdc
Collector Current — Continuous	I_C	200	mAdc
Total Power Dissipation @ T_A = 25°C Derate above 25°C	P_D	350 2.8	mW mW/°C
Total Power Dissipation @ T_C = 25°C Derate above 25°C	P_D	1.0 8.0	Watts mW/°C
Junction Operating Temperature	T_J	150	°C
Storage Temperature Range	T_{stg}	–55 to +150	°C

THERMAL CHARACTERISTICS

Characteristic	Symbol	Max	Unit
Thermal Resistance, Junction to Ambient	$R_{\theta JA}$	357	°C/W
Thermal Resistance, Junction to Case	$R_{\theta JC}$	125	°C/W

*Indicates JEDEC Registered Data

STYLE 1
PIN 1.   EMITTER
     2.   BASE
     3.   COLLECTOR

DIM	MILLIMETERS		INCHES	
	MIN	MAX	MIN	MAX
A	4.450	5.200	0.175	0.205
B	3.180	4.190	0.125	0.165
C	4.320	5.330	0.170	0.210
D	0.407	0.533	0.016	0.021
F	0.407	0.482	0.016	0.019
K	12.700	–	0.500	–
L	1.150	1.390	0.045	0.055
N	–	1.270	–	0.050
P	6.350	–	0.250	–
Q	3.430	–	0.135	–
R	2.410	2.670	0.095	0.105
S	2.030	2.670	0.080	0.105

CASE 29-02
TO-92

*ELECTRICAL CHARACTERISTICS (T$_A$ = 25°C unless otherwise noted)

Characteristic		Fig. No.	Symbol	Min	Max	Unit
OFF CHARACTERISTICS						
Collector-Base Breakdown Voltage (I_C = 10 μAdc, I_E = 0)			BV_{CBO}	60	-	Vdc
Collector-Emitter Breakdown Voltage (1) (I_C = 1.0 mAdc, I_B = 0)			BV_{CEO}	40	-	Vdc
Emitter-Base Breakdown Voltage (I_E = 10 μAdc, I_C = 0)			BV_{EBO}	6.0	-	Vdc
Collector Cutoff Current (V_{CE} = 30 Vdc, $V_{EB(off)}$ = 3.0 Vdc)			I_{CEV}	-	50	nAdc
Base Cutoff Current (V_{CE} = 30 Vdc, $V_{EB(off)}$ = 3.0 Vdc)			I_{BEV}	-	50	nAdc
ON CHARACTERISTICS						
DC Current Gain (1) (I_C = 0.1 mAdc, V_{CE} = 1.0 Vdc)	2N3903 2N3904	15	h_{FE}	20 40	- -	-
(I_C = 1.0 mAdc, V_{CE} = 1.0 Vdc)	2N3903 2N3904			35 70	- -	
(I_C = 10 mAdc, V_{CE} = 1.0 Vdc)	2N3903 2N3904			50 100	150 300	
(I_C = 50 mAdc, V_{CE} = 1.0 Vdc)	2N3903 2N3904			30 60	- -	
(I_C = 100 mAdc, V_{CE} = 1.0 Vdc)	2N3903 2N3904			15 30	- -	
Collector-Emitter Saturation Voltage (1) (I_C = 10 mAdc, I_B = 1.0 mAdc) (I_C = 50 mAdc, I_B = 5.0 mAdc)		16, 17	$V_{CE(sat)}$	- -	0.2 0.3	Vdc
Base-Emitter Saturation Voltage (1) (I_C = 10 mAdc, I_B = 1.0 mAdc) (I_C = 50 mAdc, I_B = 5.0 mAdc)		17	$V_{BE(sat)}$	0.65 -	0.85 0.95	Vdc
SMALL-SIGNAL CHARACTERISTICS						
Current-Gain—Bandwidth Product (I_C = 10 mAdc, V_{CE} = 20 Vdc, f = 100 MHz)	2N3903 2N3904		f_T	250 300	- -	MHz
Output Capacitance (V_{CB} = 5.0 Vdc, I_E = 0, f = 100 kHz)		3	C_{ob}	-	4.0	pF
Input Capacitance (V_{BE} = 0.5 Vdc, I_C = 0, f = 100 kHz)		3	C_{ib}	-	8.0	pF
Input Impedance (I_C = 1.0 mAdc, V_{CE} = 10 Vdc, f = 1.0 kHz)	2N3903 2N3904	13	h_{ie}	0.5 1.0	8.0 10	k ohms
Voltage Feedback Ratio (I_C = 1.0 mAdc, V_{CE} = 10 Vdc, f = 1.0 kHz)	2N3903 2N3904	14	h_{re}	0.1 0.5	5.0 8.0	X 10-4
Small-Signal Current Gain (I_C = 1.0 mAdc, V_{CE} = 10 Vdc, f = 1.0 kHz)	2N3903 2N3904	11	h_{fe}	50 100	200 400	-
Output Admittance (I_C = 1.0 mAdc, V_{CE} = 10 Vdc, f = 1.0 kHz)		12	h_{oe}	1.0	40	μmhos
Noise Figure (I_C = 100 μAdc, V_{CE} = 5.0 Vdc, R_S = 1.0 k ohms, f = 10 Hz to 15.7 kHz)	2N3903 2N3904	9, 10	NF	- -	6.0 5.0	dB
SWITCHING CHARACTERISTICS						
Delay Time (V_{CC} = 3.0 Vdc, $V_{BE(off)}$ = 0.5 Vdc, I_C = 10 mAdc, I_{B1} = 1.0 mAdc)		1, 5	t_d	-	35	ns
Rise Time		1, 5, 6	t_r	-	35	ns
Storage Time (V_{CC} = 3.0 Vdc, I_C = 10 mAdc, I_{B1} = I_{B2} = 1.0 mAdc)	2N3903 2N3904	2, 7	t_s	- -	175 200	ns
Fall Time		2, 8	t_f	-	50	ns

(1) Pulse Test: Pulse Width = 300 μs, Duty Cycle = 2.0%.
*Indicates JEDEC Registered Data

FIGURE 1 — DELAY AND RISE TIME EQUIVALENT TEST CIRCUIT

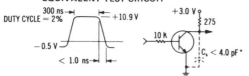

FIGURE 2 — STORAGE AND FALL TIME EQUIVALENT TEST CIRCUIT

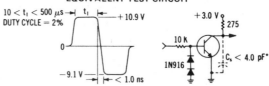

*Total shunt capacitance of test jig and connectors

MOTOROLA
Semiconductors

BOX 20912 • PHOENIX, ARIZONA 85036

2N3905
2N3906

PNP SILICON ANNULAR♦ TRANSISTORS

.... designed for general purpose switching and amplifier applications and for complementary circuitry with types 2N3903 and 2N3904.

- High Voltage Ratings — BV_{CEO} = 40 Volts (Min)
- Current Gain Specified from 100 μA to 100 mA
- Complete Switching and Amplifier Specifications
- Low Capacitance — C_{ob} = 4.5 pF (Max)

PNP SILICON
SWITCHING & AMPLIFIER
TRANSISTORS

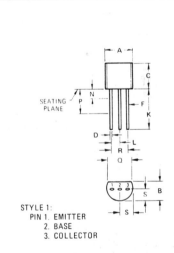

STYLE 1:
PIN 1. EMITTER
2. BASE
3. COLLECTOR

*MAXIMUM RATINGS

Rating	Symbol	Value	Unit
Collector-Base Voltage	V_{CB}	40	Vdc
Collector-Emitter Voltage	V_{CEO}	40	Vdc
Emitter-Base Voltage	V_{EB}	5.0	Vdc
Collector Current	I_C	200	mAdc
Total Power Dissipation @ T_A = 60°C	P_D	250	mW
Total Power Dissipation @ T_A = 25°C Derate above 25°C	P_D	350 2.8	mW mW/°C
Total Power Dissipation @ T_C = 25°C Derate above 25°C	P_D	1.0 8.0	Watt mW/°C
Junction Operating Temperature	T_J	+150	°C
Storage Temperature Range	T_{stg}	−55 to +150	°C

THERMAL CHARACTERISTICS

Characteristic	Symbol	Max	Unit
Thermal Resistance, Junction to Ambient	$R_{\theta JA}$	357	°C/W
Thermal Resistance, Junction to Case	$R_{\theta JC}$	125	°C/W

	MILLIMETERS		INCHES	
DIM	MIN	MAX	MIN	MAX
A	4.450	5.200	0.175	0.205
B	3.180	4.190	0.125	0.165
C	4.320	5.330	0.170	0.210
D	0.407	0.533	0.016	0.021
F	0.407	0.482	0.016	0.019
K	12.700	—	0.500	—
L	1.150	1.390	0.045	0.055
N	—	1.270	—	0.060
P	6.350	—	0.250	—
Q	3.430	—	0.135	—
R	2.410	2.670	0.095	0.105
S	2.030	2.670	0.080	0.105

CASE 29-02
(TO-92)

*Indicates JEDEC Registered Data.
♦Annular semiconductors patented by Motorola Inc.

DS 5128 R2

Characteristic		Fig. No.	Symbol	Min	Max	Unit
OFF CHARACTERISTICS						
Collector-Base Breakdown Voltage (I_C = 10 μAdc, I_E = 0)			BV_{CBO}	40	—	Vdc
Collector-Emitter Breakdown Voltage (1) (I_C = 1.0 mAdc, I_B = 0)			BV_{CEO}	40	—	Vdc
Emitter-Base Breakdown Voltage (I_E = 10 μAdc, I_C = 0)			BV_{EBO}	5.0	—	Vdc
Collector Cutoff Current (V_{CE} = 30 Vdc, $V_{BE(off)}$ = 3.0 Vdc)			I_{CEX}	—	50	nAdc
Base Cutoff Current (V_{CE} = 30 Vdc, $V_{BE(off)}$ = 3.0 Vdc)			I_{BL}	—	50	nAdc
ON CHARACTERISTICS (1)						
DC Current Gain			h_{FE}			
(I_C = 0.1 mAdc, V_{CE} = 1.0 Vdc)	2N3905	15		30	—	
	2N3906			60	—	
(I_C = 1.0 mAdc, V_{CE} = 1.0 Vdc)	2N3905			40	—	
	2N3906			80	—	
(I_C = 10 mAdc, V_{CE} = 1.0 Vdc)	2N3905			50	150	
	2N3906			100	300	
(I_C = 50 mAdc, V_{CE} = 1.0 Vdc)	2N3905			30	—	
	2N3906			60	—	
(I_C = 100 mAdc, V_{CE} = 1.0 Vdc)	2N3905			15	—	
	2N3906			30	—	
Collector-Emitter Saturation Voltage		16, 17	$V_{CE(sat)}$			Vdc
(I_C = 10 mAdc, I_B = 1.0 mAdc)				—	0.25	
(I_C = 50 mAdc, I_B = 5.0 mAdc)				—	0.4	
Base-Emitter Saturation Voltage		17	$V_{BE(sat)}$			Vdc
(I_C = 10 mAdc, I_B = 1.0 mAdc)				0.65	0.85	
(I_C = 50 mAdc, I_B = 5.0 mAdc)				—	0.95	
SMALL-SIGNAL CHARACTERISTICS						
Current-Gain — Bandwidth Product			f_T			MHz
(I_C = 10 mAdc, V_{CE} = 20 Vdc, f = 100 MHz)	2N3905			200	—	
	2N3906			250	—	
Output Capacitance (V_{CB} = 5.0 Vdc, I_E = 0, f = 100 kHz)		3	C_{ob}	—	4.5	pF
Input Capacitance (V_{BE} = 0.5 Vdc, I_C = 0, f = 100 kHz)		3	C_{ib}	—	1.0	pF
Input Impedance		13	h_{ie}			k ohms
(I_C = 1.0 mAdc, V_{CE} = 10 Vdc, f = 1.0 kHz)	2N3905			0.5	8.0	
	2N3906			2.0	12	
Voltage Feedback Ratio		14	h_{re}			X 10^{-4}
(I_C = 1.0 mAdc, V_{CE} = 10 Vdc, f = 1.0 kHz)	2N3905			0.1	5.0	
	2N3906			1.0	10	
Small-Signal Current Gain		11	h_{fe}			—
(I_C = 1.0 mAdc, V_{CE} = 10 Vdc, f = 1.0 kHz)	2N3905			50	200	
	2N3906			100	400	
Output Admittance		12	h_{oe}			μmhos
(I_C = 1.0 mAdc, V_{CE} = 10 Vdc, f = 1.0 kHz)	2N3905			1.0	40	
	2N3906			3.0	60	
Noise Figure		9, 10	NF			dB
(I_C = 100 μAdc, V_{CE} = 5.0 Vdc, R_S = 1.0 k ohm,	2N3905			—	5.0	
f = 10 Hz to 15.7 kHz)	2N3906			—	4.0	
SWITCHING CHARACTERISTICS						
Delay Time	(V_{CC} = 3.0 Vdc, $V_{BE(off)}$ = 0.5 Vdc	1, 5	t_d	—	35	ns
Rise Time	I_C = 10 mAdc, I_{B1} = 1.0 mAdc)	1, 5, 6	t_r	—	35	ns
Storage Time	2N3905	2, 7	t_s	—	200	ns
	(V_{CC} = 3.0 Vdc, I_C = 10 mAdc, 2N3906			—	225	
Fall Time	I_{B1} = I_{B2} = 1.0 mAdc) 2N3905	2, 8	t_f	—	60	ns
	2N3906			—	75	

*Indicates JEDEC Registered Data. (1) Pulse Width = 300 μs, Duty Cycle = 2.0 %.

FIGURE 1 — DELAY AND RISE TIME EQUIVALENT TEST CIRCUIT

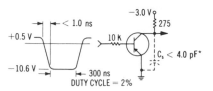

FIGURE 2 — STORAGE AND FALL TIME EQUIVALENT TEST CIRCUIT

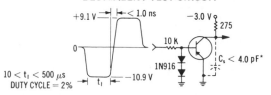

*Total shunt capacitance of test jig and connectors

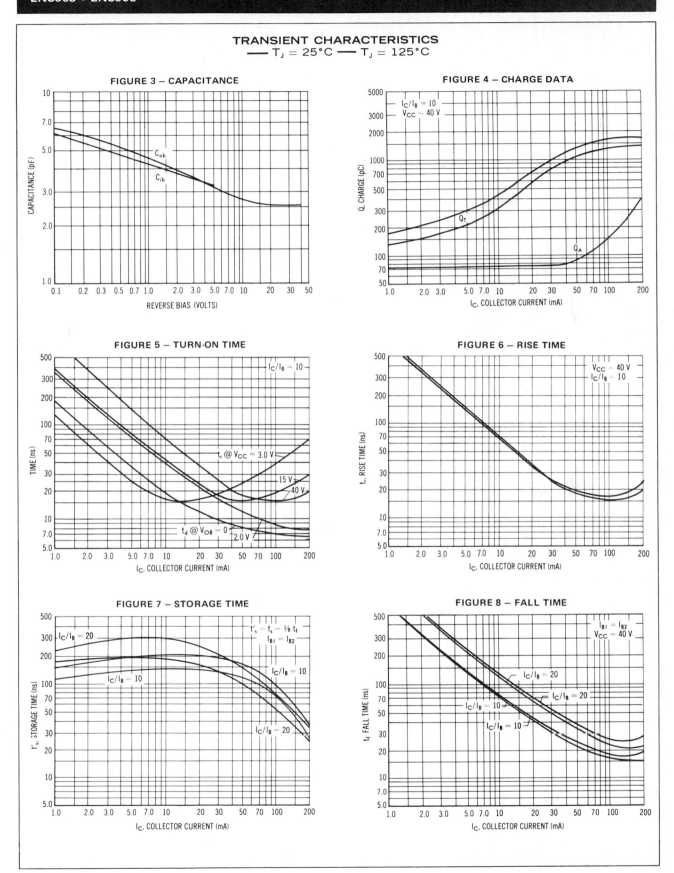

TRANSIENT CHARACTERISTICS
—— $T_J = 25°C$ —— $T_J = 125°C$

FIGURE 3 — CAPACITANCE

FIGURE 4 — CHARGE DATA

FIGURE 5 — TURN-ON TIME

FIGURE 6 — RISE TIME

FIGURE 7 — STORAGE TIME

FIGURE 8 — FALL TIME

2N4918 thru 2N4920 (SILICON)
MJE4918 thru MJE4920

MEDIUM-POWER PLASTIC PNP SILICON TRANSISTORS

. . . designed for driver circuits, switching, and amplifier applications. These high-performance plastic devices feature:

- Low Saturation Voltage — $V_{CE(sat)} = 0.6$ Vdc (Max) @ $I_C = 1.0$ Amp

- Excellent Power Dissipation Due to Thermopad Construction — $P_D = 30$ and 40 W @ $T_C = 25^oC$

- Excellent Safe Operating Area

- Gain Specified to $I_C = 1.0$ Amp

- Complement to NPN 2N4921, 2N4922, 2N4923 and MJE4921, MJE4922, MJE4923

- Choice of Packages — 2N4918 thru 2N4920, 30 Watts, Case 77
  MJE4918 thru MJE4920, 40 Watts, Case 199

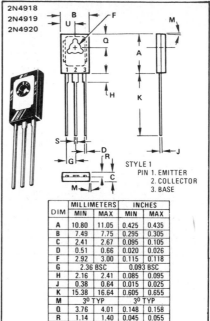

2N4918
2N4919
2N4920

STYLE 1
PIN 1. EMITTER
2. COLLECTOR
3. BASE

DIM	MILLIMETERS		INCHES	
	MIN	MAX	MIN	MAX
A	10.80	11.05	0.425	0.435
B	7.49	7.75	0.295	0.305
C	2.41	2.67	0.095	0.105
D	0.51	0.66	0.020	0.026
F	2.92	3.00	0.115	0.118
G	2.36 BSC		0.093 BSC	
H	2.16	2.41	0.085	0.095
J	0.38	0.64	0.015	0.025
K	15.38	16.64	0.605	0.655
M	3° TYP		3° TYP	
Q	3.76	4.01	0.148	0.158
R	1.14	1.40	0.045	0.055
S	0.64	0.89	0.025	0.035
U	3.68	3.94	0.145	0.155

CASE 77-03

*MAXIMUM RATINGS

Ratings	Symbol	2N4918 MJE4918	2N4919 MJE4919	2N4920 MJE4920	Unit
Collector-Emitter Voltage	V_{CEO}	40	60	80	Vdc
Collector-Base Voltage	V_{CB}	40	60	80	Vdc
Emitter-Base Voltage	V_{EB}	← 5.0 →			Vdc
Collector Current — Continuous (1)	I_C*	← 1.0 →			Adc
		← 3.0 →			
Base Current	I_B	← 1.0 →			Adc
		2N4918 series	**MJE4918 series**		
Total Device Dissipation @ $T_C = 25^oC$ Derate above 25°C	P_D	30 0.24	40 0.32		Watts W/°C
Operating & Storage Junction Temperature Range	T_J, T_{stg}	← −65 to +150 →			°C

THERMAL CHARACTERISTICS (2)

Characteristic	Symbol	2N4918/20	MJE4918/20	Unit
Thermal Resistance, Junction to Case	θ_{JC}	4.16	3.125	°C/W

*Indicates JEDEC Registered Data for 2N4918 Series
(1) The 1.0 Amp maximum I_C value is based upon JEDEC current gain requirements. The 3.0 Amp maximum value is based upon actual current-handling capability of the device (See Figure 5).
(2) Recommend use of thermal compound for lowest thermal resistance.

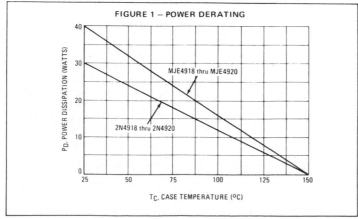

FIGURE 1 — POWER DERATING

MJE4918 thru MJE4920

2N4918 thru 2N4920

P_D, POWER DISSIPATION (WATTS)

T_C, CASE TEMPERATURE (°C)

MJE4918
MJE4919
MJE4920

STYLE 1:
PIN 1. BASE
2. COLLECTOR
3. EMITTER

DIM	MILLIMETERS		INCHES	
	MIN	MAX	MIN	MAX
A	16.08	16.33	0.633	0.643
B	12.57	12.83	0.495	0.505
C	3.18	3.43	0.125	0.135
D	0.51	0.76	0.020	0.030
F	3.61	3.86	0.142	0.152
G	2.54 BSC		0.100 BSC	
H	2.67	2.92	0.105	0.115
J	0.43	0.69	0.017	0.027
K	14.73	14.99	0.580	0.590
L	2.16	2.41	0.085	0.095
M	3° TYP		3° TYP	
N	1.47	1.73	0.058	0.068
Q	4.78	5.03	0.188	0.198
R	1.91	2.16	0.075	0.085
S	0.81	0.86	0.032	0.034
T	6.99	7.24	0.275	0.285
U	6.22	6.48	0.245	0.255

1. DIM "G" IS TO CENTER LINE OF LEADS.
CASE 199-04

ELECTRICAL CHARACTERISTICS (T_C = 25°C unless otherwise noted)

Characteristic	Fig. No.	Symbol	Min	Max	Unit
OFF CHARACTERISTICS					
Collector-Emitter Sustaining Voltage (1)	—	$V_{CEO(sus)}$		—	Vdc
(I_C = 0.1 Adc, I_B = 0) 2N4918,MJE4918			40	—	
2N4919,MJE4919			60	—	
2N4920,MJE4920			80	—	
Collector Cutoff Current	—	I_{CEO}			mAdc
(V_{CE} = 20 Vdc, I_B = 0) 2N4918,MJE4918			—	0.5	
(V_{CE} = 30 Vdc, I_B = 0) 2N4919,MJE4919			—	0.5	
(V_{CE} = 40 Vdc, I_B = 0) 2N4920,MJE4920			—	0.5	
Collector Cutoff Current	13	I_{CEX}			mAdc
(V_{CE} = Rated V_{CEO}, $V_{BE(off)}$ = 1.5 Vdc)			—	0.1	
(V_{CE} = Rated V_{CEO}, $V_{BE(off)}$ = 1.5 Vdc, T_C = 125°C)			—	0.5	
Collector Cutoff Current	—	I_{CBO}		•	mAdc
(V_{CB} = Rated V_{CB}, I_E = 0)			—	0.1	
Emitter Cutoff Current	—	I_{EBO}	—	1.0	mAdc
(V_{BE} = 5.0 Vdc, I_C = 0)					
ON CHARACTERISTICS					
DC Current Gain (1)	9	h_{FE}			—
(I_C = 50 mAdc, V_{CE} = 1.0 Vdc)			40	—	
(I_C = 500 mAdc, V_{CE} = 1.0 Vdc)			20	100	
(I_C = 1.0 Adc, V_{CE} = 1.0 Vdc)			10	—	
Collector-Emitter Saturation Voltage (1)	10	$V_{CE(sat)}$			Vdc
(I_C = 1.0 Adc, I_B = 0.1 Adc)	12		—	0.6	
	14				
Base-Emitter Saturation Voltage (1)	12	$V_{BE(sat)}$			Vdc
(I_C = 1.0 Adc, I_B = 0.1 Adc)	14		—	1.3	
Base-Emitter On Voltage (1)	12	$V_{BE(on)}$			Vdc
(I_C = 1.0 Adc, V_{CE} = 1.0 Vdc)	14		—	1.3	
SMALL-SIGNAL CHARACTERISTICS					
Current-Gain — Bandwidth Product	—	f_T			MHz
(I_C = 250 mAdc, V_{CE} = 10 Vdc, f = 1.0 MHz)			3.0	—	
Output Capacitance	—	C_{ob}			pF
(V_{CB} = 10 Vdc, I_E = 0, f = 100 kHz)			—	100	
Small-Signal Current Gain	—	h_{fe}			—
(I_C = 250 mAdc, V_{CE} = 10 Vdc, f = 1.0 kHz)			25	—	

*Indicates JEDEC Registered Data for 2N4918 Series.

(1) Pulse Test: PW $\approx$ 300 μs, Duty Cycle $\approx$ 2.0%

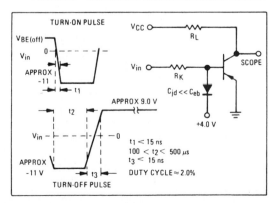

FIGURE 2 — SWITCHING TIME EQUIVALENT CIRCUIT

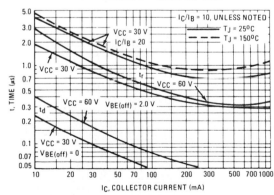

FIGURE 3 — TURN-ON TIME

2N4921 thru 2N4923 (SILICON)
MJE4921 thruMJE4923

MEDIUM-POWER PLASTIC NPN SILICON TRANSISTORS

. . . designed for driver circuits, switching, and amplifier applications. These high-performance plastic devices feature:

- Low Saturation Voltage — $V_{CE(sat)}$ = 0.6 Vdc (Max) @ I_C = 1.0 Amp
- Excellent Power Dissipation Due to Thermopad Construction — P_D = 30 and 40 W @ T_C = 25°C
- Excellent Safe Operating Area
- Gain Specified to I_C = 1.0 Amp
- Complement to PNP 2N4918, 2N4919, 2N4920 and MJE4918, MJE4919, MJE4920
- Choice of Packages — 2N4921 thru 2N4923, 30 Watts — Case 77
  MJE4921 thru MJE4923, 40 Watts — Case 199

*MAXIMUM RATINGS

Rating	Symbol	2N4921 MJE4921	2N4922 MJE4922	2N4923 MJE4923	Unit
Collector-Emitter Voltage	V_{CEO}	40	60	80	Vdc
Collector-Base Voltage	V_{CB}	40	60	80	Vdc
Emitter-Base Voltage	V_{EB}		5.0		Vdc
Collector Current · Continuous (1)	I_C		1.0		Adc
			3.0		
Base Current — Continuous	I_B		1.0		Adc

		2N4921 Series	MJE4921 Series	
Total Device Dissipation @ T_C = 25°C	P_D	30	40	Watts
Derate above 25°C		0.24	0.32	W/°C
Operating & Storage Junction Temperature Range	T_J, T_{stg}		−65 to +150	°C

THERMAL CHARACTERISTICS (2)

Characteristic	Symbol	2N4921/4923	MJE4921/4923	Unit
Thermal Resistance, Junction to Case	θ_{JC}	4.16	3.125	°C/W

(1) The 1.0 Amp maximum I_C value is based upon JEDEC current gain requirements.
The 3.0 Amp maximum value is based upon actual current handling capability of the device (see Figures 5 and 6).
(2) Recommend use of thermal compound for lowest thermal resistance.
*Indicates JEDEC Registered Data for 2N4921 Series.

FIGURE 1 – POWER DERATING

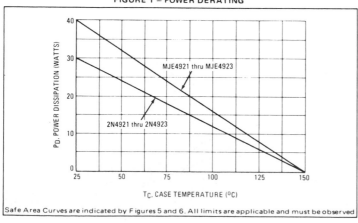

Safe Area Curves are indicated by Figures 5 and 6. All limits are applicable and must be observed

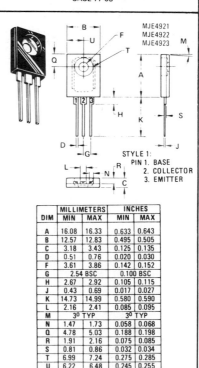

2N4921
2N4922
2N4923

STYLE 1
PIN 1. EMITTER
2. COLLECTOR
3. BASE

DIM	MILLIMETERS MIN	MAX	INCHES MIN	MAX
A	10.80	11.05	0.425	0.435
B	7.49	7.75	0.295	0.305
C	2.41	2.67	0.095	0.105
D	0.51	0.66	0.020	0.026
F	2.92	3.00	0.115	0.118
G	2.36 BSC		0.093 BSC	
H	2.16	2.41	0.085	0.095
J	0.38	0.64	0.015	0.025
K	15.38	16.64	0.605	0.655
M	3° TYP		3° TYP	
Q	3.76	4.01	0.148	0.158
R	1.14	1.40	0.045	0.055
S	0.64	0.89	0.025	0.035
U	3.68	3.94	0.145	0.155

CASE 77-03

MJE4921
MJE4922
MJE4923

STYLE 1:
PIN 1. BASE
2. COLLECTOR
3. EMITTER

DIM	MILLIMETERS MIN	MAX	INCHES MIN	MAX
A	16.08	16.33	0.633	0.643
B	12.57	12.83	0.495	0.505
C	3.18	3.43	0.125	0.135
D	0.51	0.76	0.020	0.030
F	3.61	3.86	0.142	0.152
G	2.54 BSC		0.100 BSC	
H	2.67	2.92	0.105	0.115
J	0.43	0.69	0.017	0.027
K	14.73	14.99	0.580	0.590
L	2.16	2.41	0.085	0.095
M	3° TYP		3° TYP	
N	1.47	1.73	0.058	0.068
Q	4.78	5.03	0.188	0.198
R	1.91	2.16	0.075	0.085
S	0.81	0.86	0.032	0.034
T	6.99	7.24	0.275	0.285
U	6.22	6.48	0.245	0.255

1. DIM "G" IS TO CENTER LINE OF LEADS.
CASE 199-04

2N4921 thru 2N4923, MJE4921 thru MJE4923 (continued)

*ELECTRICAL CHARACTERISTICS (T_C = 25°C unless otherwise noted)

Characteristic	Figure No.	Symbol	Min	Max	Unit
OFF CHARACTERISTICS					
Collector-Emitter Sustaining Voltage (1)	—	$V_{CEO(sus)}$			Vdc
(I_C = 0.1 Adc, I_B = 0) 2N4921, MJE4921			40	—	
2N4922, MJE4922			60	—	
2N4923, MJE4923			80	—	
Collector Cutoff Current	—	I_{CEO}			mAdc
(V_{CE} = 20 Vdc, I_B = 0) 2N4921, MJE4921			—	0.5	
(V_{CE} = 30 Vdc, I_B = 0) 2N4922, MJE4922			—	0.5	
(V_{CE} = 40 Vdc, I_B = 0) 2N4923, MJE4923			—	0.5	
Collector Cutoff Current	13	I_{CEX}			mAdc
(V_{CE} = Rated V_{CEO}, $V_{EB(off)}$ = 1.5 Vdc)			—	0.1	
(V_{CE} = Rated V_{CEO}, $V_{EB(off)}$ = 1.5 Vdc, T_C = 125°C)			—	0.5	
Collector Cutoff Current	—	I_{CBO}			mAdc
(V_{CB} = Rated V_{CB}, I_E = 0)			—	0.1	
Emitter Cutoff Current	—	I_{EBO}			mAdc
(V_{EB} = 5.0 Vdc, I_C = 0)			—	1.0	
ON CHARACTERISTICS					
DC Current Gain (1)	9	h_{FE}			—
(I_C = 50 mAdc, V_{CE} = 1.0 Vdc)			40	—	
(I_C = 500 mAdc, V_{CE} = 1.0 Vdc)			20	100	
(I_C = 1.0 Adc, V_{CE} = 1.0 Vdc)			10	—	
Collector-Emitter Saturation Voltage (1)	10	$V_{CE(sat)}$			Vdc
(I_C = 1.0 Adc, I_B = 0.1 Adc)	12		—	0.6	
	14				
Base-Emitter Saturation Voltage (1)	12	$V_{BE(sat)}$			Vdc
(I_C = 1.0 Adc, I_B = 0.1 Adc)	14		—	1.3	
Base-Emitter On Voltage (1)	12	$V_{BE(on)}$			Vdc
(I_C = 1.0 Adc, V_{CE} = 1.0 Vdc)	14		—	1.3	
SMALL-SIGNAL CHARACTERISTICS					
Current-Gain — Bandwidth Product	—	f_T			MHz
(I_C = 250 mAdc, V_{CE} = 10 Vdc, f = 1.0 MHz)			3.0	—	
Output Capacitance	—	C_{ob}			pF
(V_{CB} = 10 Vdc, I_E = 0, f = 100 kHz)			—	100	
Small-Signal Current Gain	—	h_{fe}			—
(I_C = 250 mAdc, V_{CE} = 10 Vdc, f = 1.0 kHz)			25	—	

(1) Pulse Test: PW ≈ 300 μs, Duty Cycle ≈ 2.0%.
*Indicates JEDEC Registered Data for 2N4921 Series.

FIGURE 2 — SWITCHING TIME EQUIVALENT CIRCUIT

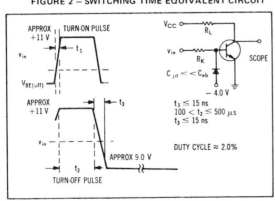

FIGURE 3 — TURN-ON TIME

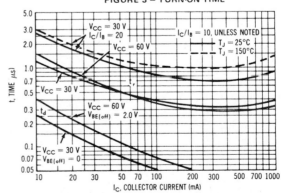

MOTOROLA
Semiconductors
BOX 20912 • PHOENIX, ARIZONA 85036

4N29, 4N29A
4N30
4N31
4N32, 4N32A
4N33

INFRARED LIGHT EMITTING DIODE PHOTO DARLINGTON TRANSISTOR COUPLED PAIR

NPN PHOTOTRANSISTOR AND PN INFRARED EMITTING DIODE

. . . Gallium Arsenide LED optically coupled to a Silicon Photo Darlington Transistor designed for applications requiring electrical isolation, high-current transfer ratios, small package size and low cost; such as interfacing and coupling systems, phase and feedback controls, solid-state relays and general-purpose switching circuits.

- High Isolation Voltage
  $V_{ISO} = 7500$ V (Min)

- High Collector Output Current
  @ $I_F = 10$ mA –
  $I_C = 50$ mA (Min) – 4N32,33
  10 mA (Min) – 4N29,30
  5.0 mA (Min) – 4N31

- Economical, Compact, Dual-In-Line Package

- Excellent Frequency Response –
  30 kHz (Typ)

- Fast Switching Times @ $I_C = 50$ mA
  $t_{on} = 0.6 \mu s$ (Typ)
  $t_{off} = 17 \mu s$ (Typ) – 4N29,30,31
  45 μs (Typ) – 4N32,33

- 4N29A, 4N32A are UL Recognized –
  File Number E54915

MAXIMUM RATINGS ($T_A = 25^oC$ unless otherwise noted)

Rating	Symbol	Value	Unit
INFRARED-EMITTING DIODE MAXIMUM RATINGS			
Reverse Voltage	V_R	3.0	Volts
Forward Current – Continuous	I_F	80	mA
Forward Current – Peak (Pulse Width = 300 μs, 2.0% Duty Cycle)	I_F	3.0	Amp
Total Power Dissipation @ $T_A = 25^oC$ Negligible Power in Transistor	P_D	150	mW
Derate above 25°C		2.0	mW/^{o}C
PHOTOTRANSISTOR MAXIMUM RATINGS			
Collector-Emitter Voltage	V_{CEO}	30	Volts
Emitter-Collector Voltage	V_{ECO}	5.0	Volts
Collector-Base Voltage	V_{CBO}	30	Volts
Total Power Dissipation @ $T_A = 25^oC$	P_D	150	mW
Negligible Power in Diode Derate above 25°C		2.0	mW/^{o}C
TOTAL DEVICE RATINGS			
Total Device Dissipation @ $T_A = 25^oC$ Equal Power Dissipation in Each Element	P_D	250	mW
Derate above 25°C		3.3	mW/^{o}C
Operating Junction Temperature Range	T_J	-55 to +100	^{o}C
Storage Temperature Range	T_{stg}	-55 to +150	^{o}C
Soldering Temperature (10 s)	–	260	^{o}C

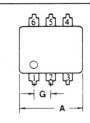

STYLE 1:
PIN 1. ANODE
2. CATHODE
3. NC
4. EMITTER
5. COLLECTOR
6. BASE

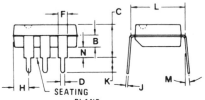

NOTES:
1. LEADS WITHIN 0.25 mm (0.010) DIAMETER OF TRUE POSITION AT SEATING PLANE AT MAXIMUM MATERIAL CONDITION.
2. DIMENSION "L" TO CENTER OF LEADS WHEN FORMED PARALLEL.

	MILLIMETERS		INCHES	
DIM	MIN	MAX	MIN	MAX
A	8.13	8.89	0.320	0.350
B	1.27	2.03	0.050	0.080
C	2.92	5.08	0.115	0.200
D	0.41	0.51	0.016	0.020
F	1.02	1.78	0.040	0.070
G	2.54	BSC	0.100	BSC
H	1.02	2.16	0.040	0.085
J	0.20	0.30	0.008	0.012
K	2.54	3.81	0.100	0.150
L	7.62	BSC	0.300	BSC
M	0°	15°	0°	15°
N	0.38	2.54	0.015	0.100

CASE 730-01

FIGURE 1 – MAXIMUM POWER DISSIPATION

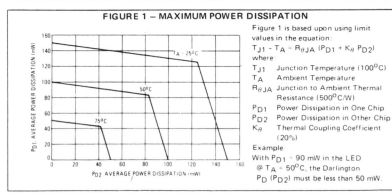

Figure 1 is based upon using limit values in the equation:
$T_{J1} - T_A = R_{\theta JA} (P_{D1} + K_\theta P_{D2})$
where:
T_{J1} Junction Temperature (100°C)
T_A Ambient Temperature
$R_{\theta JA}$ Junction to Ambient Thermal Resistance (500°C/W)
P_{D1} Power Dissipation in One Chip
P_{D2} Power Dissipation in Other Chip
K_θ Thermal Coupling Coefficient (20%)
Example:
With $P_{D1} = 90$ mW in the LED @ $T_A = 50^oC$, the Darlington P_D (P_{D2}) must be less than 50 mW.

DS 2627 R3

LED CHARACTERISTICS (T$_A$ = 25°C unless otherwise noted)

Characteristic	Symbol	Min	Typ	Max	Unit
*Reverse Leakage Current (V$_R$ = 3.0 V, R$_L$ = 1.0 M ohms)	I$_R$	—	0.05	100	μA
*Forward Voltage (I$_F$ = 50 mA)	V$_F$	—	1.2	1.5	Volts
Capacitance (V$_R$ = 0 V, f = 1.0 MHz)	C	—	150	—	pF

PHOTOTRANSISTOR CHARACTERISTICS (T$_A$ = 25°C and I$_F$ = 0 unless otherwise noted)

Characteristic	Symbol	Min	Typ	Max	Unit
*Collector-Emitter Dark Current (V$_{CE}$ = 10 V, Base Open)	I$_{CEO}$	—	—	100	nA
*Collector-Base Breakdown Voltage (I$_C$ = 100 μA, I$_E$ = 0)	BV$_{CBO}$	30	—	—	Volts
*Collector-Emitter Breakdown Voltage (I$_C$ = 100 μA, I$_B$ = 0)	BV$_{CEO}$	30	—	—	Volts
*Emitter-Collector Breakdown Voltage (I$_E$ = 100 μA, I$_B$ = 0)	BV$_{ECO}$	5.0	—	—	Volts
DC Current Gain (V$_{CE}$ = 5.0 V, I$_C$ = 500 μA)	h$_{FE}$	—	5000	—	—

COUPLED CHARACTERISTICS (T$_A$ = 25°C unless otherwise noted)

Characteristic		Symbol	Min	Typ	Max	Unit
*Collector Output Current (1)	4N32, 4N33	I$_C$	50	—	—	mA
(V$_{CE}$ = 10 V, I$_F$ = mA, I$_B$ = 0	4N29, 4N30		10	—	—	
	4N31		5.0	—	—	
Isolation Surge Voltage (2, 5)		V$_{ISO}$				Volts
(60 Hz ac Peak, 5 Seconds)			7500	—	—	
	*4N29, 4N32		2500	—	—	
	*4N30, 4N31, 4N33		1500	—	—	
Isolation Resistance (2) (V = 500 V)		—	—	10^{11}	—	Ohms
*Collector-Emitter Saturation Voltage (1)	4N31	V$_{CE(sat)}$	—	—	1.2	Volts
(I$_C$ = 2.0 mA, I$_F$ = 8.0 mA)	4N29, 4N39, 4N32, 4N33		—	—	1.0	
Isolation Capacitance (2) (V = 0, f = 1.0 MHz)		—	—	0.8	—	pF
Bandwidth (3) (I$_C$ = 2.0 mA, R$_L$ = 100 ohms, Figures 6 and 8)		—	—	30	—	kHz

SWITCHING CHARACTERISTICS (Figures 7 and 9), (4)

Characteristic		Symbol	Min	Typ	Max	Unit
Turn-On Time (I$_C$ = 50 mA, I$_F$ = 200 mA, V$_{CC}$ = 10 V)		t$_{on}$	—	0.6	5.0	μs
Turn-Off Time	4N29, 30, 31	t$_{off}$	—	17	40	μs
(I$_C$ = 50 mA, I$_F$ = 200 mA, V$_{CC}$ = 10 V)	4N32, 33		—	45	100	

*Indicates JEDEC Registered Data.
(1) Pulse Test: Pulse Width = 300 μs, Duty Cycle ≤ 2.0%.
(2) For this test, LED pins 1 and 2 are common and phototransistor pins 4, 5, and 6 are common.
(3) I$_F$ adjusted to yield I$_C$ = 2.0 mA and i$_c$ = 2.0 mA P-P at 10 kHz.
(4) t$_d$ and t$_r$ are inversely proportional to the amplitude of I$_F$; t$_s$ and t$_f$ are not significantly affected by I$_F$.
(5) Isolation Surge Voltage, V$_{ISO}$, is an internal device dielectric breakdown rating.

DC CURRENT TRANSFER CHARACTERISTICS

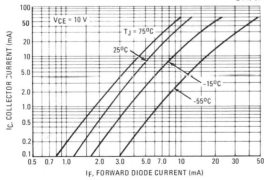

FIGURE 2 — 4N29, 4N30, 4N31

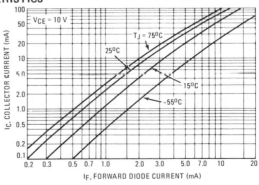

FIGURE 3 — 4N32, 4N33

MOTOROLA *Semiconductor Products Inc.*

TYPICAL ELECTRICAL CHARACTERISTICS
(Printed Circuit Board Mounting)

FIGURE 4 — DIODE FORWARD CHARACTERISTIC

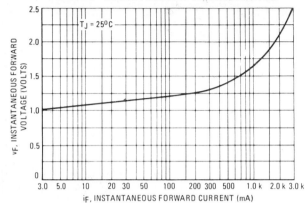

FIGURE 5 — COLLECTOR-EMITTER CUTOFF CURRENT

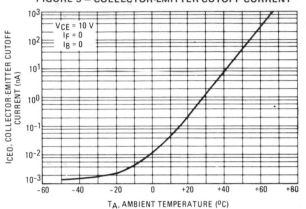

FIGURE 6 — FREQUENCY RESPONSE

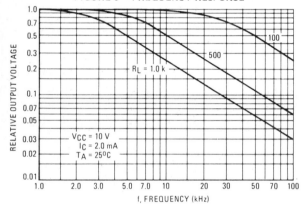

FIGURE 7 — SWITCHING TIMES

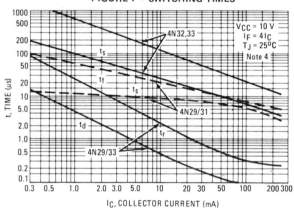

FIGURE 8 — FREQUENCY RESPONSE TEST CIRCUIT

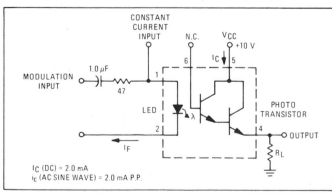

FIGURE 9 — SWITCHING TIME TEST CIRCUIT

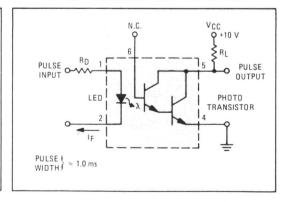

Ⓜ **MOTOROLA** *Semiconductor Products Inc.*

MOTOROLA

SEMICONDUCTORS

P.O. BOX 20912 • PHOENIX, ARIZONA 85036

**4N35
4N36
4N37**

NPN PHOTOTRANSISTOR AND PN INFRARED EMITTING DIODE

. . . gallium-arsenide LED optically coupled to a silicon photo-transistor designed for applications requiring electrical isolation, high-current transfer ratios, small package size and low cost such as interfacing and coupling systems, phase and feedback controls, solid-state relays and general-purpose switching circuits.

- High Electrical Isolation V_{ISO} = 7500 V (Min)
- High Transfer Ratio —
  100% (min) @ I_F = 10 mA, V_{CE} = 10 V
- Low Collector-Emitter Saturation Voltage —
  $V_{CE(sat)}$ = 0.3 Vdc (max) @ I_F = 10 mA, I_C = 0.5 mA
- UL Recognized File Number E54915

OPTO COUPLER

PHOTOTRANSISTOR OUTPUT

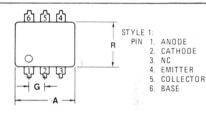

STYLE 1:
PIN 1. ANODE
2. CATHODE
3. NC
4. EMITTER
5. COLLECTOR
6. BASE

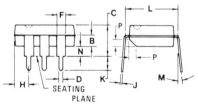

NOTES:
1. LEADS WITHIN 0.25 mm (0.010) DIAMETER OF TRUE POSITION AT SEATING PLANE AT MAXIMUM MATERIAL CONDITION.
2. DIMENSION "L" TO CENTER OF LEADS WHEN FORMED PARALLEL.

MAXIMUM RATINGS (T_A = 25°C unless otherwise noted)

Rating	Symbol	Value	Unit
*INFRARED-EMITTER DIODE MAXIMUM RATINGS			
Reverse Voltage	V_{RB}	6.0	Volts
Forward Current — Continuous	I_F	60	mA
Forward Current — Peak Pulse Width = 1.0 μs, 2.0% Duty Cycle	I_F	3.0	Amp
Total Power Dissipation @ T_A = 25°C Negligible Power in Transistor Derate above 25°C	P_D	 100 1.3	mW mW/°C
Total Power Dissipation @ T_C = 25°C Derate above 25°C	P_D	100 1.3	mW mW/°C
*PHOTOTRANSISTOR MAXIMUM RATINGS			
Collector-Emitter Voltage	V_{CEO}	30	Volts
Emitter-Base Voltage	V_{EBO}	7.0	Volts
Collector-Base Voltage	V_{CBO}	70	Volts
Output Current — Continuous	I_C	100	mA
Total Power Dissipation @ T_A = 25°C Negligible Power in Diode Derate above 25°C	P_D	 300 4.0	mW mW/°C
Total Power Dissipation @ T_C = 25°C Derate above 25°C	P_D	500 6.7	mW mW/°C
TOTAL DEVICE RATINGS			
*Total Power Dissipation @ T_A = 25°C Derate above 25°C	P_D	300 3.3	mW mW/°C
Input to Output Isolation Voltage, Surge 60 Hz Peak ac, 5 seconds JEDEC Registered 4N35 = 3500 V Data @ 8 ms 4N36 = 2500 V 4N37 = 1500 V	V_{ISO}	7500	Volts V_{pk}
*Junction Temperature Range	T_J	–55 to +100	°C
*Storage Temperature Range	T_{stg}	–55 to +150	°C
*Soldering Temperature (10 s)	—	260	°C

*Indicates JEDEC Registered Data

DIM	MILLIMETERS		INCHES	
	MIN	MAX	MIN	MAX
A	8.13	8.89	0.320	0.350
B	1.27	2.03	0.050	0.080
C	2.92	5.08	0.115	0.200
D	0.41	0.51	0.016	0.020
F	1.02	1.78	0.040	0.070
G	2.54	BSC	0.100	BSC
H	1.02	2.16	0.040	0.085
J	0.20	0.30	0.008	0.012
K	2.54	3.81	0.100	0.150
L	7.62	BSC	0.300	BSC
M	0°	15°	0°	15°
N	0.38	2.54	0.015	0.100
P	0.81	0.97	0.032	0.038
R	6.10	6.60	0.240	0.260

CASE 730-01

DS 2649

ELECTRICAL CHARACTERISTICS

Characteristic	Symbol	Min	Typ	Max	Unit
LED CHARACTERISTICS ($T_A = 25^oC$ unless otherwise noted)					
*Reverse Leakage Current ($V_R = 6.0$ V)	I_R	—	0.05	10	μA
*Forward Voltage ($I_F = 10$ mA) ($I_F = 10$ mA, $T_A = -55^oC$) ($I_F = 10$ mA, $T_A = 100^oC$)	V_F	 0.8 0.9 0.7	 1.2 — —	 1.5 1.7 1.4	Volts
Capacitance ($V_R = 0$ V, f = 1.0 MHz)	C	—	150	—	pF
***PHOTOTRANSISTOR CHARACTERISTICS** ($T_A = 25^oC$ and $I_F = 0$ unless otherwise noted)					
Collector-Emitter Dark Current ($V_{CE} = 10$ V, Base Open) ($V_{CE} = 30$ V, Base Open, $T_A = 100^oC$)	I_{CEO}	 — —	 3.5 —	 50 500	 nA μA
Collector-Base Dark Current ($V_{CB} = 10$ V, Emitter Open)	I_{CBO}	—	—	20	nA
Collector-Base Breakdown Voltage ($I_C = 100 \mu A$, $I_E = 0$)	BV_{CBO}	70	—	—	Volts
Collector-Emitter Breakdown Voltage ($I_C = 1.0$ mA, $I_B = 0$)	BV_{CEO}	30	—	—	Volts
Emitter-Base Breakdown Voltage ($I_E = 100 \mu A$, $I_B = 0$)	BV_{EBO}	7.0	—	—	Volts
***COUPLED CHARACTERISTICS** ($T_A = 25^oC$ unless otherwise noted)					
Current Transfer Ratio ($V_{CE} = 10$ V, $I_F = 10$ mA) ($V_{CE} = 10$ V, $I_F = 10$ mA, $T_A = -55^oC$) ($V_{CE} = 10$ V, $I_F = 10$ mA, $T_A = 100^oC$)	I_C/I_F	 1.0 0.4 0.4	 — — —	 — — —	—
Input to Output Isolation Current (2) (3) ($V_{io} = 3550$ V$_{pk}$) 4N35 ($V_{io} = 2500$ V$_{pk}$) 4N36 ($V_{io} = 1500$ V$_{pk}$) 4N37	I_{IO}	 — — —	 — — —	 100 100 100	μA
Isolation Resistance (2) (V = 500 V)	R_{IO}	10^{11}	—	—	Ohms
Collector-Emitter Saturation Voltage ($I_C = 0.5$ mA, $I_F = 10$ mA)	$V_{CE(sat)}$	—	—	0.3	Volts
Isolation Capacitance (2) (V = 0, f = 1.0 MHz)	—	—	1.3	2.5	pF
***SWITCHING CHARACTERISTICS** (Figure 1)					
Turn-On Time ($V_{CC} = 10$ V, $I_C = 2.0$ mA, $R_L = 100 \Omega$)	t_{on}	—	—	10	μs
Turn-Off Time ($V_{CC} = 10$ V, $I_C = 2.0$ mA, $R_L = 100 \Omega$)	t_{off}	—	—	10	μs

* Indicates JEDEC Registered Data.

NOTES: 1. Pulse Test. Pulse Width = 300 μs, Duty Cycle ≤ 2.0%.
      2. For this test LED pins 1 and 2 are common and phototransistor pins 4, 5, and 6 are common.
      3. Pulse Width ≤ 8.0 ms.

 MOTOROLA *Semiconductor Products Inc.*

TYPICAL ELECTRICAL CHARACTERISTICS

FIGURE 1 — SWITCHING TIMES TEST CIRCUIT

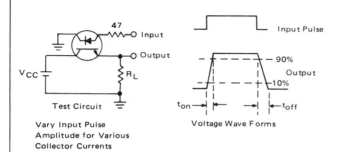

Test Circuit

Vary Input Pulse
Amplitude for Various
Collector Currents

Voltage Wave Forms

FIGURE 2 — DIODE FORWARD CHARACTERISTICS

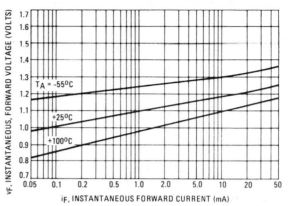

FIGURE 3 — COLLECTOR SATURATION REGION

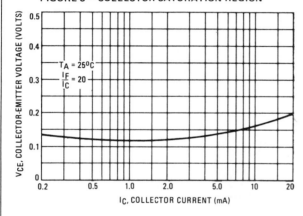

FIGURE 4 — COLLECTOR BASE CURRENT
versus INPUT CURRENT

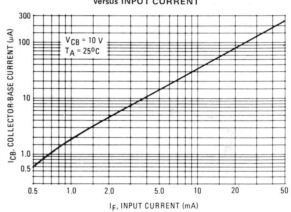

FIGURE 5 — COLLECTOR LEAKAGE CURRENT
versus TEMPERATURE

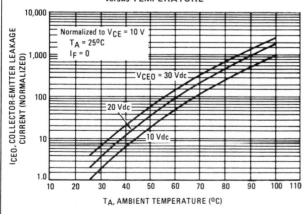

FIGURE 6 — COLLECTOR CHARACTERISTICS

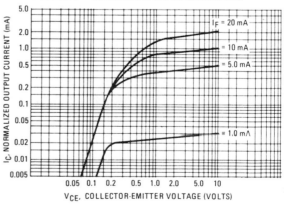

 MOTOROLA *Semiconductor Products Inc.*

MC1733
MC1733C

DIFFERENTIAL VIDEO AMPLIFIER

. . . a wideband amplifier with differential input and differential output. Gain is fixed at 10, 100, or 400 without external components or, with the addition of one external resistor, gain becomes adjustable from 10 to 400.

- Bandwidth — 120 MHz typical @ $A_{vd} = 10$
- Rise Time — 2.5 ns typical @ $A_{vd} = 10$
- Propagation Delay Time — 3.6 ns typical @ $A_{vd} = 10$

DIFFERENTIAL VIDEO WIDEBAND AMPLIFIER

SILICON MONOLITHIC INTEGRATED CIRCUIT

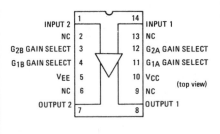

G SUFFIX
METAL PACKAGE
CASE 603
TO-100

L SUFFIX
CERAMIC PACKAGE
CASE 632
TO-116

FIGURE 1 — BASIC CIRCUIT

FIGURE 2 — VOLTAGE GAIN ADJUST CIRCUIT

CONNECTION DIAGRAMS

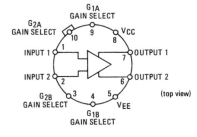

G SUFFIX, METAL PACKAGE
Pin 5 connected to case.

INPUT 2	1		14	INPUT 1
NC	2		13	NC
G2B GAIN SELECT	3		12	G2A GAIN SELECT
G1B GAIN SELECT	4		11	G1A GAIN SELECT
VEE	5		10	VCC
NC	6		9	NC
OUTPUT 2	7		8	OUTPUT 1

(top view)

L SUFFIX, CERAMIC PACKAGE

FIGURE 3 — EQUIVALENT CIRCUIT SCHEMATIC

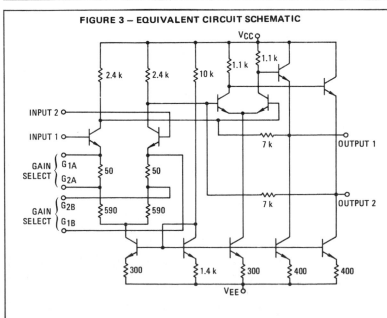

DS 9177 R1

MAXIMUM RATINGS (T_A = +25°C unless otherwise noted)

Rating	Symbol	Value	Unit
Power Supply Voltage	V_{CC} V_{EE}	+8.0 −8.0	Volts
Differential Input Voltage	V_{in}	±5.0	Volts
Common-Mode Input Voltage	V_{ICM}	±6.0	Volts
Output Current	I_O	10	mA

Internal Power Dissipation (Note 1)
  Metal Can Package
  Ceramic Dual In-Line Package
	P_D	
		500
		500
			mW

Operating Temperature Range MC1733C MC1733	T_A	0 to +70 −55 to +125	°C
Storage Temperature Range	T_{stg}	−65 to +150	°C

ELECTRICAL CHARACTERISTICS (V_{CC} = +6.0 Vdc, V_{EE} = −6.0 Vdc, at T_A = +25°C unless otherwise noted.)

Characteristic	Symbol	MC1733 Min	MC1733 Typ	MC1733 Max	MC1733C Min	MC1733C Typ	MC1733C Max	Units

Differential Voltage Gain
  Gain 1 (Note 2)
  Gain 2 (Note 3)
  Gain 3 (Note 4)
	A_{vd}	
		300
		90
		9.0
			
			400
			100
			10
				
				500
				110
				11
					
					250
					80
					8.0
						
						400
						100
						10
							
							600
							120
							12
								V/V

Bandwidth   (R_s = 50 Ω)
  Gain 1
  Gain 2
  Gain 3
	BW	
		—
		—
		—
			
			40
			90
			120
				
				—
				—
				—
					
					—
					—
					—
						
						40
						90
						120
							
							—
							—
							—
								MHz

Rise Time   (R_s = 50 Ω, V_O = 1 Vp-p)
  Gain 1
  Gain 2
  Gain 3
	t_{TLH}
	t_{THL}
		
		—
		—
		—
			
			10.5
			4.5
			2.5
				
				—
				10
				—
					
					—
					—
					—
						
						10.5
						4.5
						2.5
							
							—
							12
							—
								ns

Propagation Delay (R_s = 50 Ω, V_O = 1 Vp-p)
  Gain 1
  Gain 2
  Gain 3
	t_{PLH}
	t_{PHL}
		
		—
		—
		—
			
			7.5
			6.0
			3.6
				
				—
				10
				—
					
					—
					—
					—
						
						7.5
						6.0
						3.6
							
							—
							10
							—
								ns

Input Resistance
  Gain 1
  Gain 2
  Gain 3
	R_{in}	
		—
		20
		—
			
			4.0
			30
			250
				
				—
				—
				—
					
					—
					10
					—
						
						4.0
						30
						250
							
							—
							—
							—
								kΩ

Input Capacitance (Gain 2)	C_{in}	—	2.0	—	—	2.0	—	pF
Input Offset Current (Gain 3)	$|I_{IO}|$	—	0.4	3.0	—	0.4	5.0	μA
Input Bias Current (Gain 3)	I_{IB}	—	9.0	20	—	9.0	30	μA
Input Noise Voltage (R_s = 50 Ω, BW = 1 kHz to 10 MHz)	V_n	—	12	—	—	12	—	μV(rms)
Input Voltage Range (Gain 2)	V_{in}	±1.0	—	—	±1.0	—	—	V

Common-Mode Rejection Ratio
  Gain 2   (V_{CM} = ±1 V, f ≤ 100 kHz)
  Gain 2   (V_{CM} = ±1 V, f = 5 MHz)
	CMRR	
		60
		—
			
			86
			60
				
				—
				—
					
					60
					—
						
						86
						60
							
							—
							—
								dB

Supply Voltage Rejection Ratio Gain 2 (ΔV_S = ±0.5 V)	PSRR	 50	 70	 —	 50	 70	 —	dB

Output Offset Voltage
  Gain 1
  Gain 2 and Gain 3
	V_{OO}	
		—
		—
			
			0.6
			0.35
				
				1.5
				1.0
					
					—
					—
						
						0.6
						0.35
							
							1.5
							1.5
								V

Output Common-Mode Voltage (Gain3)	V_{CMO}	2.4	2.9	3.4	2.4	2.9	3.4	V
Output Voltage Swing (Gain 2)	V_O	3.0	4.0	—	3.0	4.0	—	Vp-p
Output Sink Current (Gain 2)	I_O	2.5	3.6	—	2.5	3.6	—	mA
Output Resistance	R_{out}	—	20	—	—	20	—	Ω
Power Supply Current (Gain 2)	I_D	—	18	24	—	18	24	mA

 MOTOROLA *Semiconductor Products Inc.*

ELECTRICAL CHARACTERISTICS (V_{CC} = +6.0 Vdc, V_{EE} = –6.0 Vdc, at T_A = T_{high} to T_{low} unless otherwise noted.) *

Characteristic	Symbol	MC1733 Min	MC1733 Typ	MC1733 Max	MC1733C Min	MC1733C Typ	MC1733C Max	Units
Differential Voltage Gain	A_{vd}							V/V
Gain 1 (Note 2)		200	–	600	250	–	600	
Gain 2 (Note 3)		80	–	120	80	–	120	
Gain 3 (Note 4)		8.0	–	12	8.0	–	12	
Input Resistance Gain 2	R_{in}	8.0	–	–	8.0	–	–	kΩ
Input Offset Current (Gain 3)	$|I_{IO}|$	–	–	5.0	–	–	6.0	μA
Input Bias Current (Gain 3)	I_{IB}	–	–	40	–	–	40	μA
Input Voltage Range (Gain 2)	V_{in}	±1.0	–	–	±1.0	–	–	V
Common-Mode Rejection Ratio Gain 2 (V_{CM} = ±1 V, f ≤ 100 kHz)	CMRR	50	–	–	50	–	–	dB
Supply Voltage Rejection Ratio Gain 2 (ΔV_s = ±0.5 V)	PSRR	50	–	–	50	–	–	dB
Output Offset Voltage	V_{OO}							V
Gain 1		–	–	1.5	–	–	1.5	
Gain 2 and Gain 3		–	–	1.2	–	–	1.5	
Output Voltage Swing (Gain 2)	V_O	2.5	–	–	2.5	–	–	Vp-p
Output Sink Current (Gain 2)	I_O	2.2	–	–	2.5	–	–	mA
Power Supply Current (Gain 2)	I_D	–	–	27	–	–	27	mA

*T_{low} = 0°C for MC1733C, –55°C for MC1733
T_{high} = +70°C for MC1733C, +125°C for MC1733.

NOTES

Note 1: Derate metal package at 6.5 mW/°C for operation at ambient temperatures above 75°C and dual in-line package at 9 mW/°C for operation at ambient temperatures above 100°C (see Figure 4). If operation at high ambient temperatures is required (MC1733) a heatsink may be necessary to limit maximum junction temperature to 150°C. Thermal resistance, junction-to-case, for the metal package is 69.4°C per Watt.

Note 2: Gain Select pins G_{1A} and G_{1B} connected together.

Note 3: Gain Select pins G_{2A} and G_{2B} connected together.

Note 4: All Gain Select pins open.

FIGURE 4 — MAXIMUM ALLOWABLE POWER DISSIPATION

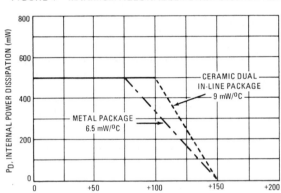

TYPICAL CHARACTERISTICS

(V_{CC} = +6.0 Vdc, V_{EE} = –6.0 Vdc, T_A = +25°C unless otherwise noted.)

FIGURE 5 — SUPPLY CURRENT versus TEMPERATURE

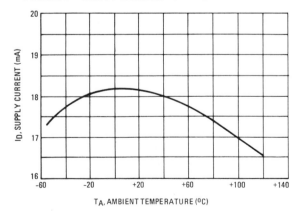

FIGURE 6 — SUPPLY CURRENT versus SUPPLY VOLTAGE

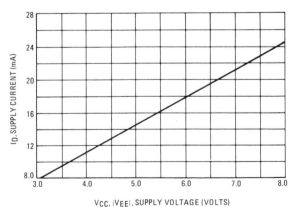

MOTOROLA *Semiconductor Products Inc.*

 **MOTOROLA**

ANALOG MULTIPLEXERS/DEMULTIPLEXERS

The MC14051B, MC14052B, and MC14053B, analog multiplexers are digitally controlled analog switches. The MC14051B effectively implements an SP8T electronic switch, the MC14052B a 2P4T, and the MC14053B a triple SPDT. All three devices feature low ON impedance and very low OFF leakage current. Control of analog signals up to the complete supply voltage range can be achieved.

- High On/Off Output Voltage Ratio — 65 dB typical
- Quiescent Current = 5.0 nA/package typical @ 5 Vdc
- Low Crosstalk Between Switches — 80 dB typical
- Diode Protection on All Inputs
- Supply Voltage Range = 3.0 Vdc to 18 Vdc
- Transmits Frequencies Up to 65 MHz
- Linearized Transfer Characteristics, $R_{ON} < 60~\Omega$ for $V_{in} = V_{DD}$ to V_{EE} @ 15 Vdc
- Low Noise — 12 nV/$\sqrt{\text{Cycle}}$, f $\geq$ 1 kHz typical
- Pin-for-Pin Replacement for CD4051, CD4052, and CD4053

CMOS MSI

(LOW-POWER COMPLEMENTARY MOS)

ANALOG MULTIPLEXERS/ DEMULTIPLEXERS

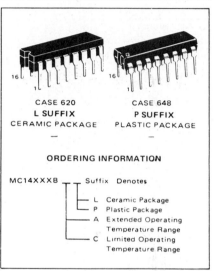

CASE 620	CASE 648
L SUFFIX	**P SUFFIX**
CERAMIC PACKAGE	PLASTIC PACKAGE
—	—

ORDERING INFORMATION

MC14XXXB ┌── Suffix Denotes

- L Ceramic Package
- P Plastic Package
- A Extended Operating Temperature Range
- C Limited Operating Temperature Range

MAXIMUM RATINGS

Rating	Symbol	Value	Unit
DC Supply Voltage	V_{DD}, V_{EE}	-0.5 to +18	Vdc
Input Voltage, All Inputs	V_{in}	-0.5 to V_{DD} + 0.5	Vdc
Through Current	I	25	mAdc
Operating Temperature Range — AL Device	T_A	-55 to +125	°C
CL/CP Device		-40 to +85	
Storage Temperature Range	T_{stg}	-65 to +150	°C

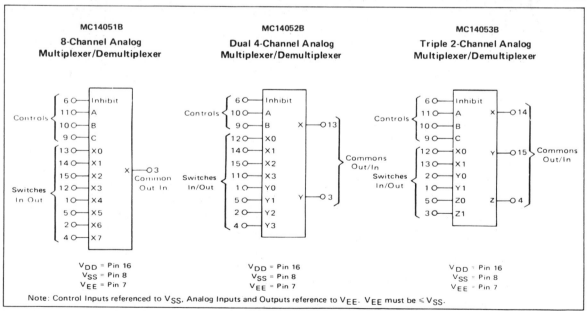

MC14051B	MC14052B	MC14053B
8-Channel Analog Multiplexer/Demultiplexer	Dual 4-Channel Analog Multiplexer/Demultiplexer	Triple 2-Channel Analog Multiplexer/Demultiplexer

V_{DD} = Pin 16
V_{SS} = Pin 8
V_{EE} = Pin 7

Note: Control Inputs referenced to V_{SS}, Analog Inputs and Outputs reference to V_{EE}. V_{EE} must be $\leq V_{SS}$.

FIGURE 1 – SWITCH CIRCUIT SCHEMATIC

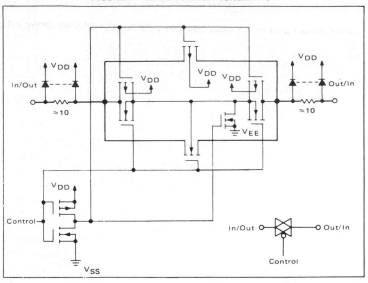

TRUTH TABLE

Control Inputs				ON Switches					
	Select								
Inhibit	C*	B	A	MC14051B	MC14052B		MC14053B		
0	0	0	0	X0	Y0	X0	Z0	Y0	X0
0	0	0	1	X1	Y1	X1	Z0	Y0	X1
0	0	1	0	X2	Y2	X2	Z0	Y1	X0
0	0	1	1	X3	Y3	X3	Z0	Y1	X1
0	1	0	0	X4			Z1	Y0	X0
0	1	0	1	X5			Z1	Y0	X1
0	1	1	0	X6			Z1	Y1	X0
0	1	1	1	X7			Z1	Y1	X1
1	x	x	x	None	None		None		

*Not applicable for MC14052
x = Don't Care

FIGURE 2 – MC14051B FUNCTIONAL DIAGRAM

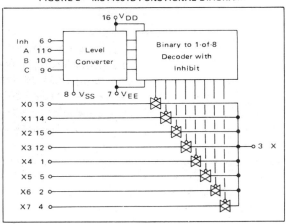

FIGURE 3 – MC14052B FUNCTIONAL DIAGRAM

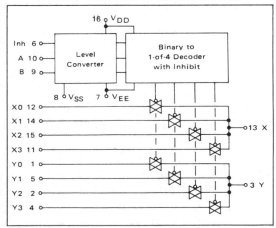

FIGURE 4 – MC14053B FUNCTIONAL DIAGRAM

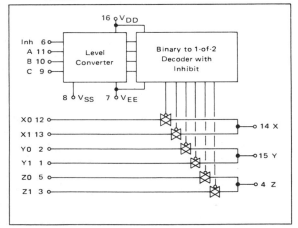

ELECTRICAL CHARACTERISTICS

Characteristic	Symbol	V_{DD}-V_{EE} Vdc	T_{low}* Min	T_{low}* Max	25°C Min	25°C Typ	25°C Max	T_{high}* Min	T_{high}* Max	Unit
Output Voltage "0" Level	V_{OL}	5.0	—	0.05	—	0	0.05	—	0.05	Vdc
V_{in} = V_{DD} or V_{SS}		10	—	0.05	—	0	0.05	—	0.05	
V_{SS} = V_{EE}		15	—	0.05	—	0	0.05	—	0.05	
"1" Level	V_{OH}	5.0	4.95	—	4.95	5.0	—	4.95	—	Vdc
V_{in} = 0 or V_{DD}		10	9.95	—	9.95	10	—	9.95	—	
		15	14.95	—	14.95	15	—	14.95	—	
Input Voltage# "0" Level	V_{IL}									Vdc
(V_O = 4.5 or 0.5 Vdc)		5.0	—	1.5	—	2.25	1.5	—	1.5	
(V_O = 9.0 or 1.0 Vdc)		10	—	3.0	—	4.50	3.0	—	3.0	
(V_O = 13.5 or 1.5 Vdc)		15	—	4.0	—	6.75	4.0	—	4.0	
"1" Level	V_{IH}									Vdc
(V_O = 0.5 or 4.5 Vdc)		5.0	3.5	—	3.5	2.75	—	3.5	—	
(V_O = 1.0 or 9.0 Vdc)		10	7.0	—	7.0	5.50	—	7.0	—	
(V_O = 1.5 or 13.5 Vdc)		15	11.25	—	11.0	8.25	—	11.0	—	
Input Current (Control, Inhibit)	I_{in}	—	—	—	—	10	—	—	—	pAdc
Input Capacitance	C_{in}									pF
(V_{in} = 0)										
Control, Inhibit		—	—	—	—	5.0	—	—	—	
Switch Inputs		—	—	—	—	10	—	—	—	
Output Capacitance MC14051B	C_{out}	10	—	—	—	60	—	—	—	pF
MC14052B		10	—	—	—	32	—	—	—	
MC14053B		10	—	—	—	17	—	—	—	
Feedthrough Capacitance MC14051B	C_{in-out}	10	—	—	—	0.18	—	—	—	pF
MC14052B		10	—	—	—	0.12	—	—	—	
MC14053B		10	—	—	—	0.10	—	—	—	
Quiescent Current (AL Device)	I_{DD}	5.0	—	5.0	—	0.005	5.0	—	150	µAdc
(Per Package)		10	—	10	—	0.010	10	—	300	
		15	—	20	—	0.015	20	—	600	
Quiescent Current (CL/CP Device)	I_{DD}	5.0	—	20	—	0.005	20	—	150	µAdc
(Per Package)		10	—	40	—	0.010	40	—	300	
		15	—	80	—	0.015	80	—	600	
Total Supply Current**†	I_T	5.0	\multicolumn{5}{} I_T = (0.07 µA/kHz) f + I_Q					µAdc
(Dynamic plus Quiescent,		10	I_T = (0.20 µA/kHz) f + I_Q							
Per Package)		15	I_T = (0.36 µA/kHz) f + I_Q							
ON Resistance (AL Device)	R_{ON}	5.0	—	880	—	250	1050	—	1200	Ω
		10	—	400	—	120	500	—	550	
		15	—	220	—	80	280	—	320	
ON Resistance (CL/CP Device)	R_{ON}	5.0	—	880	—	250	1050	—	1200	Ω
		10	—	450	—	120	500	—	520	
		15	—	250	—	80	280	—	300	
Δ ON Resistance Between Any	ΔR_{ON}	5.0	—	—	—	25	—	—	—	Ω
Two Channels		10	—	—	—	10	—	—	—	
		15	—	—	—	5.0	—	—	—	
OFF Channel Leakage Current	—									nAdc
Any Channel		15	—	100	—	±0.01	100	—	1000	
(AL Device) All Channels OFF:										
MC14051B		15	—	100	—	±0.08	100	—	1000	
MC14052B		15	—	100	—	±0.04	100	—	1000	
MC14053B		15	—	100	—	±0.02	100	—	1000	
OFF Channel Leakage Current	—									nAdc
Any Channel		15	—	1000	—	±0.01	1000	—	3000	
(CL/CP Device)										
All Channels OFF: MC10451B		15	—	1000	—	±0.08	1000	—	3000	
MC10452B		15	—	1000	—	±0.04	1000	—	3000	
MC10453B		15	—	1000	—	±0.02	1000	—	3000	

*T_{low} = -55°C for AL Device, -40°C for CL/CP Device.
T_{high} = +125°C for AL Device, +85°C for CL/CP Device.
#Noise immunity is defined as the control input voltage coincident with the specified change, ΔV_{out}, at an output in the OFF state.
**The formulas given are for the typical characteristics only at 25°C.
†Total Supply Current, I_T, is the current drawn at device terminals V_{DD} and V_{SS} for total current through the device. The channel component, (V_{in}-V_{out})/R_{ON}, should not be included.

SD 4324 SCHMITT RECEIVER

ELECTRO-OPTICAL CHARACTERISTICS (T_c=25°C UNLESS OTHERWISE SPECIFIED)

Parameter	Test Condition	Symbol	Min	Typ	Max	Units
Input Sensitivity	λp=820nm	Pmin			2	μW
Field-of-view	Note 1	FoV		40		Degrees
High Level Logic Output Voltage	P_{in}=10μW V_{cc}=5V I_{ol}>100μA	V_{oh}	2.4			volts
Low Level Logic Output Voltage	P_{in}=.5μW V_{cc}=5V I_{ol}=-16mA	V_{ol}			.4	volts
Logic Output Propagation Delay Time Low-to-High High-to-Low	V_{cc}=5VDC R_L=390Ω	td (L-H) td (H-L)		1.1 3.0		μS μS
Logic Output Transition Time Low-to-High High-to-Low		t_r t_f		80 10		ns ns
Supply Current		I_{cc}		14		mA
Operating Temperature			−40		100	°C

MOTOROLA
Semiconductors
BOX 20912 • PHOENIX, ARIZONA 85036

MC55107 MC75107
MC55108 MC75108

DUAL LINE RECEIVERS

The MC55107/MC75107 and MC55108/MC75108 are MTTL compatible dual line receivers featuring independent channels with common voltage supply and ground terminals. The MC55107/MC75107 circuit features an active pull-up (totem-pole) output. The MC55108/MC75108 circuit features an open-collector output configuration that permits the Wired-OR logic connection with similar outputs (such as the MC5401/MC7401 MTTL gate or additional MC55108/MC75108 receivers). Thus a level of logic is implemented without extra delay.

The MC55107/MC75107 and MC55108/MC75108 circuits are designed to detect input signals of greater than 25 millivolts amplitude and convert the polarity of the signal into appropriate MTTL compatible output logic levels.

- High Common-Mode Rejection Ratio
- High Input Impedance
- High Input Sensitivity
- Differential Input Common-Mode Voltage Range of ± 3.0 V
- Differential Input Common-Mode Voltage of More Than ± 15 V Using External Attenuator
- Strobe Inputs for Receiver Selection
- Gate Inputs for Logic Versatility
- MTTL or MDTL Drive Capability
- High DC Noise Margins

DUAL LINE RECEIVERS

SILICON MONOLITHIC
INTEGRATED CIRCUITS

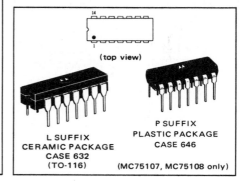

(top view)

L SUFFIX
CERAMIC PACKAGE
CASE 632
(TO-116)

P SUFFIX
PLASTIC PACKAGE
CASE 646

(MC75107, MC75108 only)

CIRCUIT SCHEMATIC

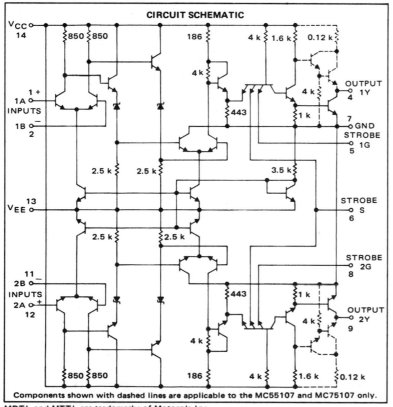

Components shown with dashed lines are applicable to the MC55107 and MC75107 only.

MDTL and MTTL are trademarks of Motorola Inc.

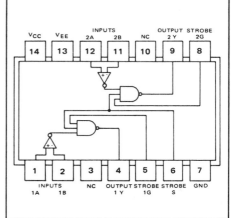

TRUTH TABLE

DIFFERENTIAL INPUTS A-B	STROBES		OUTPUT Y
	G	S	
$V_{ID} \geqslant 25$ mV	L or H	L or H	H
	L or H	L	H
-25 mV $< V_{ID} < 25$ mV	L	L or H	H
	H	H	INDETERMINATE
$V_{ID} \leqslant -25$ mV	L or H	L	H
	L	L or H	H
	H	H	L

© MOTOROLA INC., 1974   DS 9192 R1

185

MAXIMUM RATINGS ($T_A = T_{low}^*$ to T_{high}^* unless otherwise noted)

Rating	Symbol	Value	Unit
Power Supply Voltages	V_{CC} V_{EE}	+7.0 -7.0	Vdc
Differential-Mode Input Signal Voltage Range	V_{ID}	±6.0	Vdc
Common-Mode Input Voltage Range	V_{ICR}	±5.0	Vdc
Strobe Input Voltage	$V_{I(S)}$	5.5	Vdc
Power Dissipation (Package Limitation)	P_D		
Plastic and Ceramic Dual-In-Line Packages Derate above T_A = +25°C		625 3.85	mW mW/°C
Operating Ambient Temperature Range MC55107, MC55108 MC75107, MC75108	T_A	 -55 to +125 0 to +70	°C
Storage Temperature Range	T_{stg}	-65 to +150	°C

RECOMMENDED OPERATING CONDITIONS

Characteristic	Symbol	MC55107, MC55108			MC75107, MC75108			Unit
		Min	Typ	Max	Min	Typ	Max	
Power Supply Voltages	V_{CC} V_{EE}	+4.5 -4.5	+5.0 -5.0	+5.5 -5.5	+4.75 -4.75	+5.0 -5.0	+5.25 -5.25	Vdc
Output Sink Current	I_{OS}	–	–	-16	–	–	-16	mA
Differential-Mode Input Voltage Range	V_{IDR}	-5.0	–	+6.0	-5.0	–	+5.0	Vdc
Common-Mode Input Voltage Range	V_{ICR}	-3.0	–	+3.0	-3.0	–	+3.0	Vdc
Input Voltage Range, any differential input to ground	V_{IR}	-5.0	–	+3.0	-5.0	–	+3.0	Vdc
Operating Temperature Range	T_A	-55	–	+125	0	–	+70	°C

DEFINITIONS OF INPUT LOGIC LEVELS

Characteristic	Symbol	Test Fig.	Min	Max	Unit
High-Level Input Voltage (between differential inputs)	V_{IDH}	1	0.025	5.0	Vdc
Low-Level Input Voltage (between differential inputs)	V_{IDL}	1	-5.0†	-0.025	Vdc
High-Level Input Voltage (at strobe inputs)	$V_{IH(S)}$	3	2.0	5.5	Vdc
Low-Level Input Voltage (at strobe inputs)	$V_{IL(S)}$	3	0	0.8	Vdc

†The algebraic convention, where the most positive limit is designated maximum, is used with Low-Level Input Voltage Level (V_{IDL}).

ELECTRICAL CHARACTERISTICS ($T_A = T_{low}^*$ to T_{high}^* unless otherwise noted)

Characteristic	Symbol	Test Fig.	MC55107,MC75107			MC55108,MC75108			Unit
			Min	Typ #	Max	Min	Typ #	Max	
High-Level Input Current to 1A or 2A Input (V_{CC} = Max, V_{EE} = Max, V_{ID} = 0.5 V, V_{IC} = -3.0 V to +3.0 V) ‡	I_{IH}	2	–	30	75	–	30	75	μA
Low-Level Input Current to 1A or 2A Input (V_{CC} = Max, V_{EE} = Max, V_{ID} = -2.0 V, V_{IC} = -3.0 V to +3.0 V) ‡	I_{IL}	2	–	–	-10	–	–	-10	μA
High-Level Input Current to 1G or 2G Input (V_{CC} = Max, V_{EE} = Max, $V_{IH(S)}$ = 2.4 V) ‡ (V_{CC} = Max, V_{EE} = Max, $V_{IH(S)}$ = V_{CC} Max) ‡	I_{IH}	4	– –	– –	40 1.0	– –	– –	40 1.0	μA mA
Low-Level Input Current to 1G or 2G Input (V_{CC} = Max, V_{EE} = Max, $V_{IL(S)}$ = 0.4 V) ‡	I_{IL}	4	–	–	-1.6	–	–	-1.6	mA
High-Level Input Current to S Input (V_{CC} = Max, V_{EE} = Max, $V_{IH(S)}$ = 2.4 V) ‡ (V_{CC} = Max, V_{EE} = Max, $V_{IH(S)}$ = V_{CC} Max) ‡	I_{IH}	4	– –	– –	80 2.0	– –	– –	80 2.0	μA mA
Low-Level Input Current to S Input (V_{CC} = Max, V_{EE} = Max, $V_{IL(S)}$ = 0.4 V) ‡	I_{IL}	4	–	–	-3.2	–	–	-3.2	mA
High-Level Output Voltage (V_{CC} = Min, V_{EE} = Min, I_{load} = -400 μA, V_{IC} = -3.0 V to +3.0 V) ‡	V_{OH}	3	2.4	–	–	–	–	–	V
Low-Level Output Voltage (V_{CC} = Min, V_{EE} = Min, I_{sink} = 16 mA V_{IC} = -3.0 V to +3.0 V) ‡	V_{OL}	3	–	–	0.4	–	–	0.4	V
High-Level Leakage Current (V_{CC} = Min, V_{EE} = Min, V_{OH} = V_{CC} Max) ‡	I_{CEX}	3	–	–	–	–	–	250	μA
Short-Circuit Output Current ## (V_{CC} = Max, V_{EE} = Max) ‡	I_{OSC}	5	-18	–	-70	–	–	–	mA
High Logic Level Supply Current from V_{CC} (V_{CC} = Max, V_{EE} = Max, V_{ID} = 25 mV, T_A = +25°C) ‡	I_{CCH+}	6	–	18	30	–	18	30	mA
High Logic Level Supply Current from V_{EE} (V_{CC} = Max, V_{EE} = Max, V_{ID} = 25 mV, T_A = +25°C) ‡	I_{CCH-}	6	0	-8.4	-15	0	8.4	-15	mA

‡ For conditions shown as Min or Max, use the appropriate value specified under recommended operating conditions for the applicable device type.
# All typical values are at V_{CC} = +5.0 V, V_{EE} = -5.0 V, T_A = +25°C.
Not more than one output should be shorted at a time.
* T_{low} = 55°C for MC55107 and MC55108, T_{high} = +125°C for MC55107 and MC55108
   = 0 for MC75107 and MC75108   = +70°C for MC75107 and MC75108

MOTOROLA *Semiconductor Products Inc.*

186

SWITCHING CHARACTERISTICS (V_{CC} = +5.0 V, V_{EE} = −5.0 V, T_A = +25°C)

Characteristic	Symbol	Test Fig.	MC55107, MC75107			MC55108, MC75108			Unit
			Min	Typ	Max	Min	Typ	Max	
Propagation Delay Time, low-to-high level from differential inputs A and B to output (R_L = 390 Ω, C_L = 50 pF) (R_L = 390 Ω, C_L = 15 pF)	$t_{PLH(D)}$	7	− −	17 −	25 −	− −	− 19	− 25	ns
Propagation Delay Time, high-to-low level from differential inputs A and B to output (R_L = 390 Ω, C_L = 50 pF) (R_L = 390 Ω, C_L = 15 pF)	$t_{PHL(D)}$	7	− −	17 −	25 −	− −	− 19	− 25	ns
Propagation Delay Time, low-to-high level, from strobe input G or S to output (R_L = 390 Ω, C_L = 50 pF) (R_L = 390 Ω, C_L = 15 pF)	$t_{PLH(S)}$	7	− −	10 −	15 −	− −	− 13	− 20	ns
Propagation Delay Time, high-to-low level, from strobe input G or S to output (R_L = 390 Ω, C_L = 50 pF) (R_L = 390 Ω, C_L = 15 pF)	$t_{PHL(S)}$	7	− −	8.0 −	15 −	− −	− 13	− 20	ns

Symbols conform to JEDEC Bulletin No. 1 when applicable.

TEST CIRCUITS

FIGURE 1 – V_{IDH} and V_{IDL}

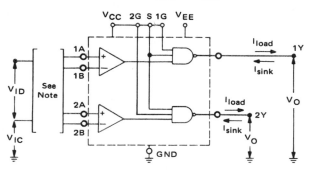

NOTE: When testing one channel, the inputs of the other channel are grounded.

FIGURE 2 – I_{IH} and I_{IL}

NOTE: Each pair of differential inputs is tested separately. The inputs of the other pair are grounded

FIGURE 3 – $V_{IH(S)}$, $V_{IL(S)}$, V_{OH}, V_{OL}, and I_{OH}

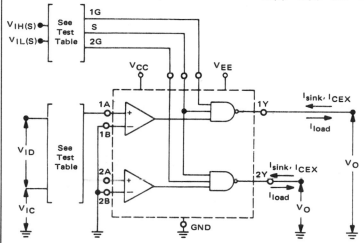

TEST TABLE

MC55107 MC75107	MC55108 MC75108	V_{ID}	STROBE 1G or 2G	STROBE S
TEST			APPLY	
V_{OH}	I_{CEX}	+25 mV	$V_{IH(S)}$	$V_{IH(S)}$
V_{OH}	I_{CEX}	−25 mV	$V_{IL(S)}$	$V_{IH(S)}$
V_{OH}	I_{CEX}	−25 mV	$V_{IH(S)}$	$V_{IL(S)}$
V_{OL}	V_{OL}	−25 mV	$V_{IH(S)}$	$V_{IH(S)}$

NOTES: 1. V_{IC} = −3.0 V to +3.0 V.
2. When testing one channel, the inputs of the other channel should be grounded.

 MOTOROLA *Semiconductor Products Inc.*

SEMICONDUCTORS

P.O. BOX 20912 • PHOENIX, ARIZONA 85036

MFOD200

PHOTOTRANSISTOR FOR FIBER OPTICS SYSTEMS

. . . designed for infrared radiation detection in medium length, medium frequency Fiber Optic Systems. Typical applications include: medical electronics, industrial controls, security systems, M6800 Microprocessor systems, etc.

- Spectral Response Matched to MFOE 200
- Hermetic Metal Package for Stability and Reliability
- High Sensitivity for Medium Length Fiber Optic Control Systems
- Compatible with AMP Mounting Bushing #227015

FIBER OPTICS
NPN SILICON
PHOTOTRANSISTOR

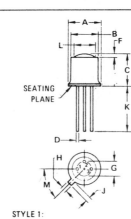

MAXIMUM RATINGS (T_A = 25°C unless otherwise noted).

Rating (Note 1)	Symbol	Value	Unit
Collector-Emitter Voltage	V_{CEO}	40	Volts
Emitter-Base Voltage	V_{EBO}	10	Volts
Collector-Base Voltage	V_{CBO}	70	Volts
Light Current	I_L	250	mA
Total Device Dissipation @ T_A = 25°C Derate above 25°C	P_D	250 1.67	mW mW/°C
Operating and Storage Junction Temperature Range	T_J, T_{stg}	−55 to +175	°C

STYLE 1:
PIN 1. EMITTER
2. BASE
3. COLLECTOR

NOTES:
1. LEADS WITHIN .13 mm (.005) RADIUS OF TRUE POSITION AT SEATING PLANE, AT MAXIMUM MATERIAL CONDITION.
2. PIN 3 INTERNALLY CONNECTED TO CASE.

DIM	MILLIMETERS		INCHES	
	MIN	MAX	MIN	MAX
A	5.31	5.84	0.209	0.230
B	4.52	4.95	0.178	0.195
C	6.22	6.98	0.245	0.275
D	0.41	0.48	0.016	0.019
F	1.19	1.60	0.047	0.063
G	2.54 BSC		0.100 BSC	
H	0.99	1.17	0.039	0.046
J	0.84	1.22	0.033	0.048
K	12.70	–	0.500	–
L	3.35	4.01	0.132	0.158
M	45° BSC		45° BSC	

CASE 82-04

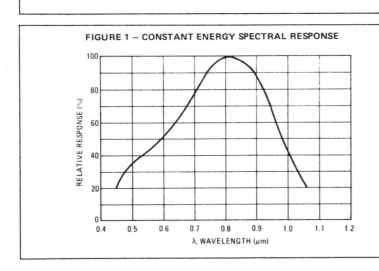

FIGURE 1 – CONSTANT ENERGY SPECTRAL RESPONSE

DS2550

STATIC ELECTRICAL CHARACTERISTICS (T_A = 25°C unless otherwise noted)

Characteristic	Symbol	Min	Typ	Max	Unit
Collector Dark Current (V_{CC} = 20 V, H ≈ 0) T_A = 25°C T_A = 100°C	I_{CEO}	– –	– 4.0	25 –	nA µA
Collector-Base Breakdown Voltage (I_C = 100 µA)	$V_{(BR)CBO}$	50	–	–	Volts
Collector-Emitter Breakdown Voltage (I_C = 100 µA)	$V_{(BR)CEO}$	30	–	–	Volts
Emitter-Collector Breakdown Voltage (I_E = 100 µA)	$V_{(BR)ECO}$	7.0	–	–	Volts

OPTICAL CHARACTERISTICS (T_A = 25°C)

Characteristic	Symbol	Min	Typ	Max	Unit
Responsivity (Figure 2)	R	14.5	18	–	µA/µW
Photo Current Rise Time (Note 1) (R_L = 100 ohms)	t_r	–	2.5	–	µs
Photo Current Fall Time (Note 1) (R_L = 100 ohms)	t_f	–	4.0	–	µs

Note 1. For unsaturated response time measurements, radiation is provided by pulsed GaAs (gallium-arsenide) light-emitting diode (λ ≈ 900 nm) with a pulse width equal to or greater than 10 microseconds, I_C = 1.0 mA peak.

FIGURE 2 – RESPONSIVITY TEST CONFIGURATION

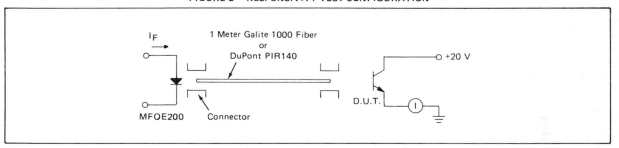

TYPICAL CHARACTERISTICS

COUPLED SYSTEM PERFORMANCE versus FIBER LENGTH*
FIGURE 3 – MFOE200 SOURCE

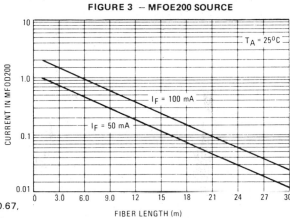

*0.045" Dia. Fiber Bundle, N.A. ≅ 0.67,
Attenuation at 900 nm ≅ 0.6 dB/m

MOTOROLA *Semiconductor Products Inc.*

BOX 20912 • PHOENIX, ARIZONA 85036 • A SUBSIDIARY OF MOTOROLA INC.

11693 4 PRINTED IN USA (4-84) MPS 3M

MFOD300

PHOTODARLINGTON TRANSISTOR FOR FIBER OPTICS SYSTEMS

. . . designed for infrared radiation detection in long length, low frequency Fiber Optics Systems. Typical applications include: industrial controls, security systems, medical electronics, M6800 Microprocessor Systems, etc.

- Spectral Response Matched to MFOE100, 200
- Hermetic Metal Package for Stability and Reliability
- Very High Sensitivity for Long Length Fiber Optics Control Systems
- Compatible With AMP Mounting Bushing #227015

FIBER OPTICS
NPN SILICON PHOTODARLINGTON TRANSISTOR

MAXIMUM RATINGS ($T_A = 25^\circ C$ unless otherwise noted).

Rating	Symbol	Value	Unit
Collector-Emitter Voltage	V_{CEO}	40	Volts
Emitter-Base Voltage	V_{EBO}	10	Volts
Collector-Base Voltage	V_{CBO}	70	Volts
Light Current	I_L	250	mA
Total Device Dissipation @ $T_A = 25^\circ C$ Derate above 25°C	P_D	250 1.67	mW mW/°C
Operating and Storage Junction Temperature Range	T_J, T_{stg}.	−55 to +175	°C

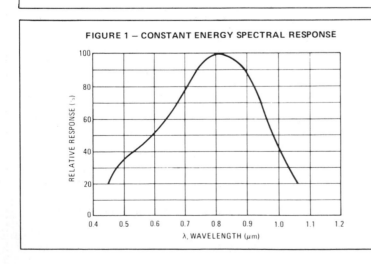

FIGURE 1 — CONSTANT ENERGY SPECTRAL RESPONSE

RELATIVE RESPONSE (%)

λ, WAVELENGTH (μm)

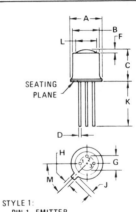

SEATING PLANE

STYLE 1:
PIN 1. EMITTER
2. BASE
3. COLLECTOR

NOTES:
1. LEADS WITHIN .13 mm (.005) RADIUS OF TRUE POSITION AT SEATING PLANE, AT MAXIMUM MATERIAL CONDITION.
2. PIN 3 INTERNALLY CONNECTED TO CASE.

	MILLIMETERS		INCHES	
DIM	MIN	MAX	MIN	MAX
A	5.31	5.84	0.209	0.230
B	4.52	4.95	0.178	0.195
C	6.22	6.98	0.245	0.275
D	0.41	0.48	0.016	0.019
F	1.19	1.60	0.047	0.063
G	2.54 BSC		0.100 BSC	
H	0.99	1.17	0.039	0.046
J	0.84	1.22	0.033	0.048
K	12.70	–	0.500	–
L	3.35	4.01	0.132	0.158
M	45° BSC		45° BSC	

CASE 82-04

DS2551
(Replacing ADI-487)

STATIC ELECTRICAL CHARACTERISTICS $(T_A = 25^\circ C)$

Characteristic	Symbol	Min	Typ	Max	Unit
Collector Dark Current $(V_{CE} = 10\ V,\ H \approx 0)$	I_{CEO}	–	10	100	nA
Collector-Base Breakdown Voltage $(I_C = 100\ \mu A)$	$V_{(BR)CBO}$	50	–	–	Volts
Collector-Emitter Breakdown Voltage $(I_C = 100\ \mu A)$	$V_{(BR)CEO}$	30	–	–	Volts
Emitter-Base Breakdown Voltage $(I_E = 100\ \mu A)$	$V_{(BR)EBO}$	7.0	–	–	Volts

OPTICAL CHARACTERISTICS $(T_A = 25^\circ C)$

Characteristic	Symbol	Min	Typ	Max	Unit
Responsivity (Figure 2)	R	400	500	–	$\mu A/\mu W$
Photo Current Rise Time (Note 1) $(R_L = 100\ ohms)$	t_r	–	40	–	μs
Photo Current Fall Time (Note 1) $(R_L = 100\ ohms)$	t_f	–	60	–	μs

Note 1. For unsaturated response time measurements, radiation is provided by pulsed GaAs (gallium-arsenide) light-emitting diode ($\lambda \approx 900$ nm) with a pulse width equal to or greater than 500 microseconds, $I_C = 1.0$ mA peak.

FIGURE 2 — RESPONSIVITY TEST CONFIGURATION

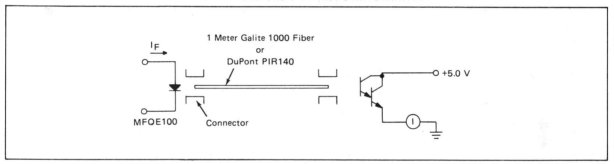

TYPICAL CHARACTERISTICS

COUPLED SYSTEM PERFORMANCE versus FIBER LENGTH*

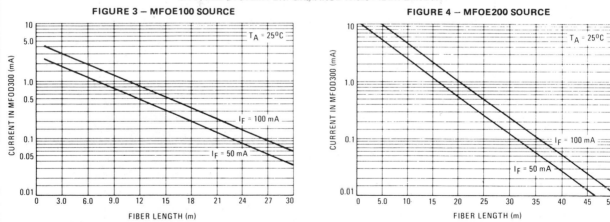

FIGURE 3 — MFOE100 SOURCE

FIGURE 4 — MFOE200 SOURCE

*0.045" Dia. Fiber Bundle, N.A. $\approx$ 0.67, Attenuation at 900 nm $\approx$ 0.6 dB/m

MOTOROLA *Semiconductor Products Inc.*

BOX 20912 • PHOENIX, ARIZONA 85036 • A SUBSIDIARY OF MOTOROLA INC.

11594-3 PRINTED IN USA 8/81 IMPERIAL LITHO 898062 2M

SEMICONDUCTORS

P.O. BOX 20912 • PHOENIX, ARIZONA 85036

MOC3009
MOC3010
MOC3011
MOC3012

OPTICALLY ISOLATED TRIAC DRIVERS

These devices consist of gallium-arsenide infrared emiting diodes, optically coupled to silicon bilateral switch and are designed for applications requiring isolated triac triggering, low-current isolated ac switching, high electrical isolation (to 7500 V peak), high detector standoff voltage, small size, and low cost.

● UL Recognized File Number 54915

OPTO
COUPLER/ISOLATOR

PHOTO TRIAC DRIVER
OUTPUT

250 VOLTS

MAXIMUM RATINGS (T_A = 25°C unless otherwise noted)

Rating	Symbol	Value	Unit
INFRARED EMITTING DIODE MAXIMUM RATINGS			
Reverse Voltage	V_R	3.0	Volts
Forward Current — Continuous	I_F	50	mA
Total Power Dissipation @ T_A = 25°C Negligible Power in Transistor Derate above 25°C	P_D	100 1.33	mW mW/°C
OUTPUT DRIVER MAXIMUM RATINGS			
Off-State Output Terminal Voltage	V_{DRM}	250	Volts
On-State RMS Current T_A = 25°C (Full Cycle , 50 to 60 Hz) T_A = 70°C	$I_{T(RMS)}$	100 50	mA mA
Peak Nonrepetitive Surge Current (PW = 10 ms, DC = 10%)	I_{TSM}	1.2	A
Total Power Dissipation @ T_A = 25°C Derate above 25°C	P_D	300 4.0	mW mW/°C
TOTAL DEVICE MAXIMUM RATINGS			
Isolation Surge Voltage (1) (Peak ac Voltage, 60 Hz, 5 Second Duration)	V_{ISO}	7500	Vac
Total Power Dissipation @ T_A = 25°C Derate above 25°C	P_D	330 4.4	mW mW/°C
Junction Temperature Range	T_J	–40 to +100	°C
Ambient Operating Temperature Range	T_A	–40 to +70	°C
Storage Temperature Range	T_{stg}	–40 to +150	°C
Soldering Temperature (10 s)	–	260	°C

(1) Isolation Surge Voltage, V_{ISO}, is an internal device dielectric breakdown rating.

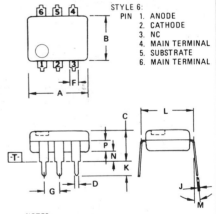

STYLE 6:
PIN 1. ANODE
2. CATHODE
3. NC
4. MAIN TERMINAL
5. SUBSTRATE
6. MAIN TERMINAL

NOTES:
1. DIMENSIONS A AND B ARE DATUMS.
2. -T- IS SEATING PLANE.
3. POSITIONAL TOLERANCES FOR LEADS:
   ⊕ ⌀ 0.13 (0.005) Ⓜ | T | AⓂBⓂ
4. DIMENSION L TO CENTER OF LEADS WHEN FORMED PARALLEL.
5. DIMENSIONING AND TOLERANCING PER ANSI Y14.5, 1973.

	MILLIMETERS		INCHES	
DIM	MIN	MAX	MIN	MAX
A	8.13	8.89	0.320	0.350
B	6.10	6.60	0.240	0.260
C	2.92	5.08	0.115	0.200
D	0.41	0.51	0.016	0.020
F	1.02	1.78	0.040	0.070
G	2.54 BSC		0.100 BSC	
J	0.20	0.30	0.008	0.012
K	2.54	3.81	0.100	0.150
L	7.62 BSC		0.300 BSC	
M	0º	15º	0º	15º
N	0.38	2.54	0.015	0.100
P	1.27	2.03	0.050	0.080

CASE 730A-01

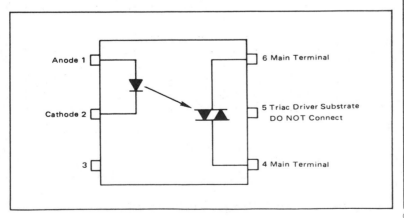

Anode 1 ——— 6 Main Terminal

Cathode 2 ——— 5 Triac Driver Substrate DO NOT Connect

3 ——— 4 Main Terminal

DS2535 R3

ELECTRICAL CHARACTERISTICS (T_A = 25°C unless otherwise noted)

Characteristic		Symbol	Min	Typ	Max	Unit
LED CHARACTERISTICS						
Reverse Leakage Current (V_R = 3.0 V)		I_R	–	0.05	100	μA
Forward Voltage (I_F = 10 mA)		V_F	–	1.2	1.5	Volts
DETECTOR CHARACTERISTICS (I_F = 0 unless otherwise noted)						
Peak Blocking Current, Either Direction (Rated V_{DRM}, Note 1)		I_{DRM}	–	10	100	nA
Peak On-State Voltage, Either Direction (I_{TM} = 100 mA Peak)		V_{TM}	–	2.5	3.0	Volts
Critical Rate of Rise of Off-State Voltage, Figure 3		dv/dt	–	2.0	–	V/μs
Critical Rate of Rise of Commutation Voltage, Figure 3 (I_{load} = 15 mA)		dv/dt	–	0.15	–	V/μs
COUPLED CHARACTERISTICS						
LED Trigger Current, Current Required to Latch Output (Main Terminal Voltage = 3.0 V)	MOC3009	I_{FT}		15	30	mA
	MOC3010		–	8.0	15	
	MOC3011		–	5.0	10	
	MOC3012		–	–	5.0	
Holding Current, Either Direction		I_H	–	100	–	μA

Note 1. Test voltage must be applied within dv/dt rating.

2. Additional information on the use of the MOC3009/3010/3011 is available in Application Note AN-780.

TYPICAL ELECTRICAL CHARACTERISTICS
T_A = 25°C

FIGURE 1 — ON-STATE CHARACTERISTICS

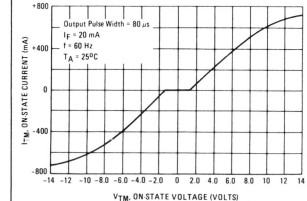

Output Pulse Width = 80 μs
I_F = 20 mA
f = 60 Hz
T_A = 25°C

I_{TM}, ON-STATE CURRENT (mA)

V_{TM}, ON-STATE VOLTAGE (VOLTS)

FIGURE 2 — TRIGGER CURRENT versus TEMPERATURE

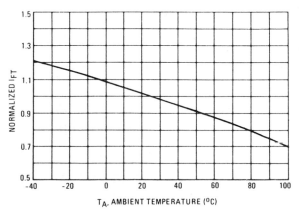

NORMALIZED I_{FT}

T_A, AMBIENT TEMPERATURE (°C)

 MOTOROLA *Semiconductor Products Inc.*

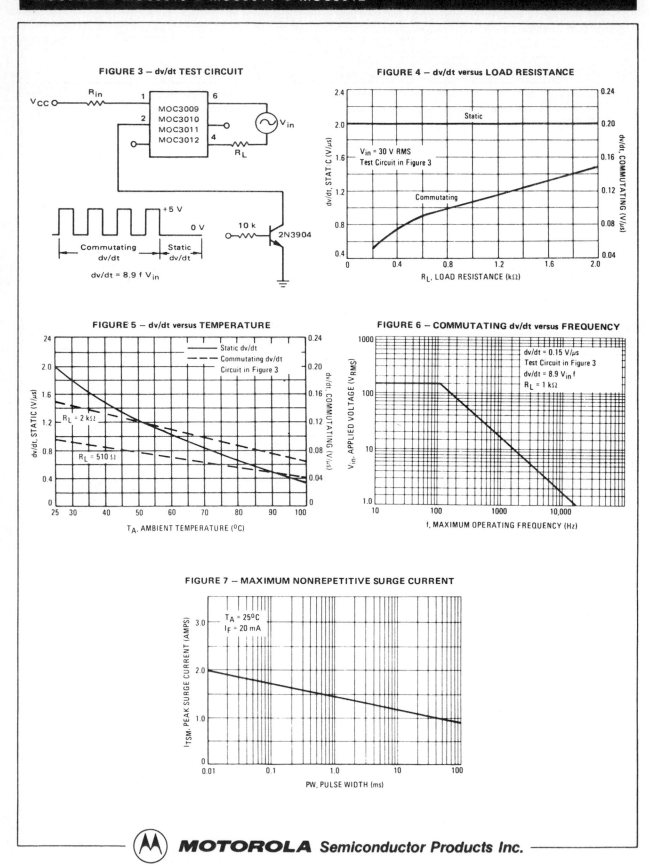

FIGURE 3 – dv/dt TEST CIRCUIT

dv/dt = 8.9 f V_{in}

FIGURE 4 – dv/dt versus LOAD RESISTANCE

V_{in} = 30 V RMS
Test Circuit in Figure 3

Static

Commutating

R_L, LOAD RESISTANCE (kΩ)

FIGURE 5 – dv/dt versus TEMPERATURE

Static dv/dt
Commutating dv/dt
Circuit in Figure 3

R_L = 2 kΩ

R_L = 510 Ω

T_A, AMBIENT TEMPERATURE (°C)

FIGURE 6 – COMMUTATING dv/dt versus FREQUENCY

dv/dt = 0.15 V/μs
Test Circuit in Figure 3
dv/dt = 8.9 V_{in} f
R_L = 1 kΩ

f, MAXIMUM OPERATING FREQUENCY (Hz)

FIGURE 7 – MAXIMUM NONREPETITIVE SURGE CURRENT

T_A = 25°C
I_F = 20 mA

PW, PULSE WIDTH (ms)

MOTOROLA *Semiconductor Products Inc.*

MOC5010

OPTICALLY ISOLATED AC LINEAR COUPLER

. . . gallium arsenide IRED optically-coupled to a bipolar monolithic amplifier. Converts an input current variation to an output voltage variation while providing a high degree of electrical isolation between input and output. Can be used for line coupling, peripheral equipment isolation, audio, medical, and other applications.

- 250 kHz Bandwidth
- Low Impedance Emitter Follower Output: $Z_O < 200\ \Omega$
- High Voltage Isolation: V_{ISO} = 7500 V (Min)
- UL Recognized, File Number E54915

OPTO COUPLER
AC LINEAR AMPLIFIER

MAXIMUM RATINGS ($T_A = 25^oC$ unless otherwise noted)

Rating	Symbol	Value	Unit
INFRARED EMITTING DIODE			
Reverse Voltage	V_R	3.0	Volts
Forward Current — Peak Pulse Width = 300 μs, 20% Duty Cycle	I_F	50	mA
Device Dissipation @ $T_A = 25^oC$ Negligible Power in IC Derate above 25^oC	P_D	100 2.0	mW mW/^oC
AC AMPLIFIER			
Supply Voltage	V_{CC}	15	Volts
Supply Current @ V_{CC} = 12 V	I_{CC}	13	mA
Device Dissipation @ $T_A = 25^oC$ Negligible Power in Diode	P_D	200	mW
TOTAL DEVICE			
Device Dissipation @ $T_A = 25^oC$	P_D	200	mW
Maximum Operating Temperature	T_A	85	^oC
Storage Temperature Range	T_{stg}	−55 to +100	^oC

FIGURE 1 — COUPLER SCHEMATIC

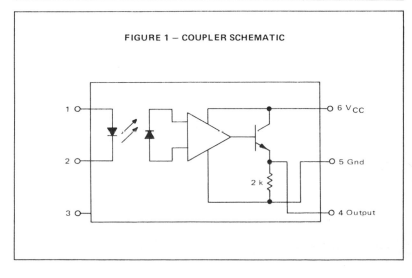

STYLE 5:
PIN 1. ANODE
     2. CATHODE
     3. NC
     4. OUTPUT
     5. GROUND
     6. V_{CC}

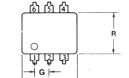

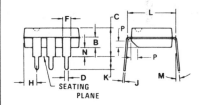

NOTES:
1. LEADS WITHIN 0.25 mm (0.010) DIAMETER OF TRUE POSITION AT SEATING PLANE AT MAXIMUM MATERIAL CONDITION.
2. DIMENSION "L" TO CENTER OF LEADS WHEN FORMED PARALLEL.

	MILLIMETERS		INCHES	
DIM	MIN	MAX	MIN	MAX
A	8.13	8.89	0.320	0.350
B	1.27	2.03	0.050	0.080
C	2.92	5.08	0.115	0.200
D	0.41	0.51	0.016	0.020
F	1.02	1.78	0.040	0.070
G	2.54	BSC	0.100	BSC
H	1.02	2.16	0.040	0.085
J	0.20	0.30	0.008	0.012
K	2.54	3.81	0.100	0.150
L	7.62	BSC	0.300	BSC
M	0°	15°	0°	15°
N	0.38	2.54	0.015	0.100
P	0.81	0.97	0.032	0.038
R	6.10	6.60	0.240	0.260

CASE 730-01

DS2542
(Replaces ADI-533)

Characteristic		Symbol	Min	Typ	Max	Unit
IRED CHARACTERISTICS ($T_A = 25°C$ unless otherwise noted)						
Reverse Leakage Current ($V_R = 3.0$ V, $R_L = 1.0$ MΩ)		I_R	–	0.05	10	μA
Forward Voltage ($I_F = 10$ mA)		V_F	–	1.2	1.5	Volts
Capacitance ($V_R = 0$ V, $f = 1.0$ MHz)		C	–	100	–	pF
ISOLATION CHARACTERISTICS ($T_A = 25°C$)						
Isolation Voltage (1) 60 Hz, AC Peak		V_{ISO}	7500	–	–	Volts
Isolation Resistance (V = 500 V) (1)		–	–	10^{11}	–	Ohms
Isolation Capacitance (V = 0, f = 1.0 MHz) (1)		–	–	1.3	–	pF
DEVICE CHARACTERISTICS ($T_A = 25°C$)						
Supply Current ($I_F = 0$, $V_{CC} = 12$ V)		I_{CC}	2.0	6.0	10	mA
Transfer Resistance — Gain	($V_{CC} = 6.0$ V)	G_R	–	100	–	mV/mA
$I_{sig} = 1.0$ mA p-p, $I_{Bias} = 12$ mA	($V_{CC} = 12$ V)		100	200	–	
Output Voltage Swing — Single Ended	($V_{CC} = 12$ V)	V_O	–	4.0	–	Volts
Single-Ended Distortion (2)		THD	See Figure 2			
Step Response		t	–	1.4	–	μs
DC Power Consumption	($V_{CC} = 6.0$ V)	P	–	30	–	mW
	($V_{CC} = 12$ V)		–	72	–	
Bandwidth		BW	100	250	–	kHz
DC Output Voltage ($I_{LED} = 0$), $V_{CE} = 12$ V		V_O	0.2	1.0	6.0	Volts

(1) For this test IRED pins 1 and 2 are common and Output Gate pins 4, 5, 6 are common.

(2) Recommended $I_F = 10$ to 15 mA at $V_{CC} = 12$ V.

FIGURE 2 — TYPICAL TOTAL HARMONIC DISTORTION

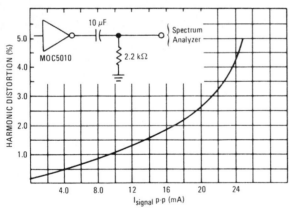

Typical total harmonic distortion @ 25°C (for units with gain of 200 mV/mA at $I_{Bias} = 12$ mA, $V_{CC} = 12$ V, f = 50 kHz, Load = [See Insert]).

FIGURE 3 — NORMALIZED FREQUENCY RESPONSE

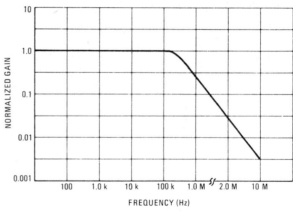

FIGURE 4 — TELEPHONE COUPLER APPLICATION

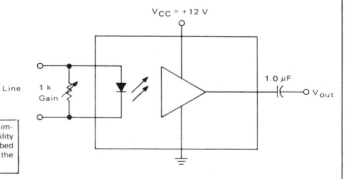

 MOTOROLA *Semiconductor Products Inc.*

BOX 20912 • PHOENIX, ARIZONA 85036 • A SUBSIDIARY OF MOTOROLA INC.

**National Semiconductor**

LM386 Low Voltage Audio Power Amplifier

General Description

The LM386 is a power amplifier designed for use in low voltage consumer applications. The gain is internally set to 20 to keep external part count low, but the addition of an external resistor and capacitor between pins 1 and 8 will increase the gain to any value up to 200.

The inputs are ground referenced while the output is automatically biased to one half the supply voltage. The quiescent power drain is only 24 milliwatts when operating from a 6 volt supply, making the LM386 ideal for battery operation.

Features

- Battery operation
- Minimum external parts
- Wide supply voltage range   4V–12V or 5V–18V
- Low quiescent current drain   4 mA

- Voltage gains from 20 to 200
- Ground referenced input
- Self-centering output quiescent voltage
- Low distortion
- Eight pin dual-in-line package

Applications

- AM-FM radio amplifiers
- Portable tape player amplifiers
- Intercoms
- TV sound systems
- Line drivers
- Ultrasonic drivers
- Small servo drivers
- Power converters

Equivalent Schematic and Connection Diagrams

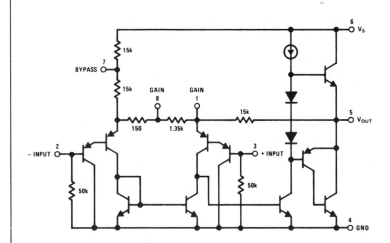

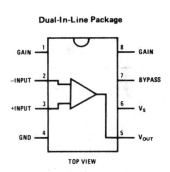

Dual-In-Line Package

TOP VIEW

Order Number LM386N-1, LM386N-3 or LM386N-4 See NS Package N08B

Typical Applications

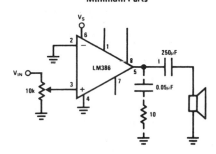

Amplifier with Gain = 20 Minimum Parts

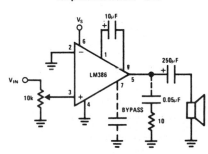

Amplifier with Gain = 200

Absolute Maximum Ratings

Supply Voltage (LM386N)	15V	Storage Temperature	-65°C to $+150^\circ$C
Supply Voltage (LM386N-4)	22V	Operating Temperature	0°C to $+70^\circ$C
Package Dissipation (Note 1) (LM386N-4)	1.25W	Junction Temperature	$+150^\circ$C
Package Dissipation (Note 2) (LM386)	660 mW	Lead Temperature (Soldering, 10 seconds)	$+300^\circ$C
Input Voltage	±0.4V		

Electrical Characteristics $T_A = 25^\circ$C

PARAMETER	CONDITIONS	MIN	TYP	MAX	UNITS
Operating Supply Voltage (V_S)					
LM386		4		12	V
LM386N-4		5		18	V
Quiescent Current (I_Q)	V_S = 6V, V_{IN} = 0		4	8	mA
Output Power (P_{OUT})					
LM386N-1	V_S = 6V, R_L = 8Ω, THD = 10%	250	325		mW
LM386N-3	V_S = 9V, R_L = 8Ω, THD = 10%	500	700		mW
LM386N-4	V_S = 16V, R_L = 32Ω, THD = 10%	700	1000		mW
Voltage Gain (A_V)	V_S = 6V, f = 1 kHz		26		dB
	10μF from Pin 1 to 8		46		dB
Bandwidth (BW)	V_S = 6V, Pins 1 and 8 Open		300		kHz
Total Harmonic Distortion (THD)	V_S = 6V, R_L = 8Ω, P_{OUT} = 125 mW f = 1 kHz, Pins 1 and 8 Open		0.2		%
Power Supply Rejection Ratio (PSRR)	V_S = 6V, f = 1 kHz, C_{BYPASS} = 10μF Pins 1 and 8 Open, Referred to Output		50		dB
Input Resistance (R_{IN})			50		kΩ
Input Bias Current (I_{BIAS})	V_S = 6V, Pins 2 and 3 Open		250		nA

Note 1: For operation in ambient temperatures above 25°C, the device must be derated based on a 150°C maximum junction temperature and a thermal resistance of 100°C/W junction to ambient.

Note 2: For operation in ambient temperatures above 25°C, the device must be derated based on a 150°C maximum junction temperature and a thermal resistance of 187°C junction to ambient.

Application Hints

GAIN CONTROL

To make the LM386 a more versatile amplifier, two pins (1 and 8) are provided for gain control. With pins 1 and 8 open the 1.35 kΩ resistor sets the gain at 20 (26 dB). If a capacitor is put from pin 1 to 8, bypassing the 1.35 kΩ resistor, the gain will go up to 200 (46 dB). If a resistor is placed in series with the capacitor, the gain can be set to any value from 20 to 200. Gain control can also be done by capacitively coupling a resistor (or FET) from pin 1 to ground.

Additional external components can be placed in parallel with the internal feedback resistors to tailor the gain and frequency response for individual applications. For example, we can compensate poor speaker bass response by frequency shaping the feedback path. This is done with a series RC from pin 1 to 5 (paralleling the internal 15 kΩ resistor). For 6 dB effective bass boost: R $\cong$ 15 kΩ, the lowest value for good stable operation is R = 10 kΩ if pin 8 is open. If pins 1 and 8 are bypassed then R as low as 2 kΩ can be used. This restriction is because the amplifier is only compensated for closed-loop gains greater than 9.

INPUT BIASING

The schematic shows that both inputs are biased to ground with a 50 kΩ resistor. The base current of the input transistors is about 250 nA, so the inputs are at about 12.5 mV when left open. If the dc source resistance driving the LM386 is higher than 250 kΩ it will contribute very little additional offset (about 2.5 mV at the input, 50 mV at the output). If the dc source resistance is less than 10 kΩ, then shorting the unused input to ground will keep the offset low (about 2.5 mV at the input, 50 mV at the output). For dc source resistances between these values we can eliminate excess offset by putting a resistor from the unused input to ground, equal in value to the dc source resistance. Of course all offset problems are eliminated if the input is capacitively coupled.

When using the LM386 with higher gains (bypassing the 1.35 kΩ resistor between pins 1 and 8) it is necessary to bypass the unused input, preventing degradation of gain and possible instabilities. This is done with a 0.1μF capacitor or a short to ground depending on the dc source resistance on the driven input.

Industrial Blocks

LM555/LM555C Timer

General Description

The LM555 is a highly stable device for generating accurate time delays or oscillation. Additional terminals are provided for triggering or resetting if desired. In the time delay mode of operation, the time is precisely controlled by one external resistor and capacitor. For astable operation as an oscillator, the free running frequency and duty cycle are accurately controlled with two external resistors and one capacitor. The circuit may be triggered and reset on falling waveforms, and the output circuit can source or sink up to 200 mA or drive TTL circuits.

Features

- Direct replacement for SE555/NE555
- Timing from microseconds through hours
- Operates in both astable and monostable modes

- Adjustable duty cycle
- Output can source or sink 200 mA
- Output and supply TTL compatible
- Temperature stability better than 0.005% per °C
- Normally on and normally off output

Applications

- Precision timing
- Pulse generation
- Sequential timing
- Time delay generation
- Pulse width modulation
- Pulse position modulation
- Linear ramp generator

Schematic Diagram

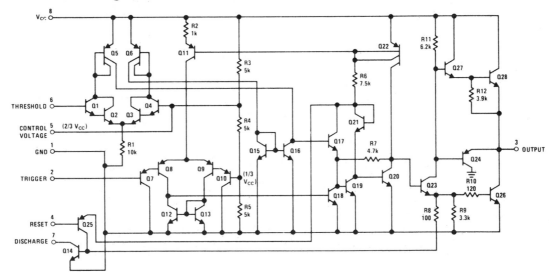

Connection Diagrams

Metal Can Package

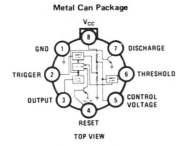

TOP VIEW

Order Number LM555H, LM555CH
See NS Package H08C

Dual-In-Line Package

1 GND	8 +Vcc
2 TRIGGER	7 DISCHARGE
3 OUTPUT	6 THRESHOLD
4 RESET	5 CONTROL VOLTAGE

TOP VIEW

Order Number LM555CN
See NS Package N08B
Order Number LM555J or LM555CJ
See NS Package J08A

Absolute Maximum Ratings

Supply Voltage	+18V
Power Dissipation (Note 1)	600 mW
Operating Temperature Ranges	
LM555C	0°C to +70°C
LM555	−55°C to +125°C
Storage Temperature Range	−65°C to +150°C
Lead Temperature (Soldering, 10 seconds)	300°C

Electrical Characteristics (T_A = 25°C, V_{CC} = +5V to +15V, unless otherwise specified)

PARAMETER	CONDITIONS	LM555 MIN	LM555 TYP	LM555 MAX	LM555C MIN	LM555C TYP	LM555C MAX	UNITS
Supply Voltage		4.5		18	4.5		16	V
Supply Current	V_{CC} = 5V, R_L = ∞		3	5		3	6	mA
	V_{CC} = 15V, R_L = ∞		10	12		10	15	mA
	(Low State) (Note 2)							
Timing Error, Monostable								
Initial Accuracy			0.5			1		%
Drift with Temperature	R_A, R_B = 1k to 100 k,		30			50		ppm/°C
	C = 0.1µF, (Note 3)							
Accuracy over Temperature			1.5			1.5		%
Drift with Supply			0.05			0.1		%/V
Timing Error, Astable								
Initial Accuracy			1.5			2.25		%
Drift with Temperature			90			150		ppm/°C
Accuracy over Temperature			2.5			3.0		%
Drift with Supply			0.15			0.30		%/V
Threshold Voltage			0.667			0.667		x V_{CC}
Trigger Voltage	V_{CC} = 15V	4.8	5	5.2		5		V
	V_{CC} = 5V	1.45	1.67	1.9		1.67		V
Trigger Current			0.01	0.5		0.5	0.9	µA
Reset Voltage		0.4	0.5	1	0.4	0.5	1	V
Reset Current			0.1	0.4		0.1	0.4	mA
Threshold Current	(Note 4)		0.1	0.25		0.1	0.25	µA
Control Voltage Level	V_{CC} = 15V	9.6	10	10.4	9	10	11	V
	V_{CC} = 5V	2.9	3.33	3.8	2.6	3.33	4	V
Pin 7 Leakage Output High			1	100		1	100	nA
Pin 7 Sat (Note 5)								
Output Low	V_{CC} = 15V, I_7 = 15 mA		150			180		mV
Output Low	V_{CC} = 4.5V, I_7 = 4.5 mA		70	100		80	200	mV
Output Voltage Drop (Low)	V_{CC} = 15V							
	I_{SINK} = 10 mA		0.1	0.15		0.1	0.25	V
	I_{SINK} = 50 mA		0.4	0.5		0.4	0.75	V
	I_{SINK} = 100 mA		2	2.2		2	2.5	V
	I_{SINK} = 200 mA		2.5			2.5		V
	V_{CC} = 5V							
	I_{SINK} = 8 mA		0.1	0.25				V
	I_{SINK} = 5 mA					0.25	0.35	V
Output Voltage Drop (High)	I_{SOURCE} = 200 mA, V_{CC} = 15V		12.5			12.5		V
	I_{SOURCE} = 100 mA, V_{CC} = 15V	13	13.3		12.75	13.3		V
	V_{CC} = 5V	3	3.3		2.75	3.3		V
Rise Time of Output			100			100		ns
Fall Time of Output			100			100		ns

Note 1: For operating at elevated temperatures the device must be derated based on a +150°C maximum junction temperature and a thermal resistance of +45°C/W junction to case for TO-5 and +150°C/W junction to ambient for both packages.

Note 2: Supply current when output high typically 1 mA less at V_{CC} = 5V.

Note 3: Tested at V_{CC} = 5V and V_{CC} = 15V.

Note 4: This will determine the maximum value of R_A + R_B for 15V operation. The maximum total (R_A + R_B) is 20 MΩ.

Note 5: No protection against excessive pin 7 current is necessary providing the package dissipation rating will not be exceeded.

Industrial Blocks

LM556/LM556C Dual Timer

General Description

The LM556 Dual timing circuit is a highly stable controller capable of producing accurate time delays or oscillation. The 556 is a dual 555. Timing is provided by an external resistor and capacitor for each timing function. The two timers operate independently of each other sharing only V_{CC} and ground. The circuits may be triggered and reset on falling waveforms. The output structures may sink or source 200 mA.

Features

- Direct replacement for SE556/NE556
- Timing from microseconds through hours
- Operates in both astable and monostable modes
- Replaces two 555 timers

- Adjustable duty cycle
- Output can source or sink 200 mA
- Output and supply TTL compatible
- Temperature stability better than 0.005% per °C
- Normally on and normally off output

Applications

- Precision timing
- Pulse generation
- Sequential timing
- Time delay generation
- Pulse width modulation
- Pulse position modulation
- Linear ramp generator

Schematic Diagram

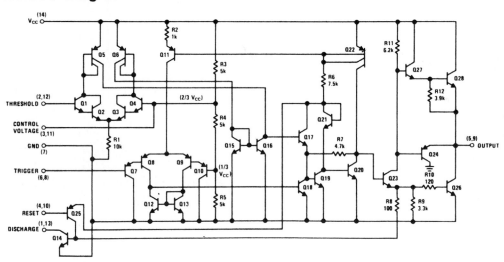

Connection Diagram

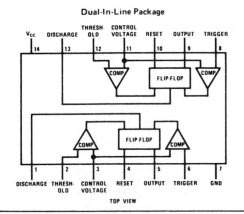

Dual-In-Line Package

TOP VIEW

Order Number LM556CN
See NS Package N14A

Order Number LM556J or LM556CJ
See NS Package J14A

Absolute Maximum Ratings

Supply Voltage	+18V
Power Dissipation (Note 1)	600 mW
Operating Temperature Ranges	
LM556C	0°C to +70°C
LM556	−55°C to +125°C
Storage Temperature Range	−65°C to +150°C
Lead Temperature (Soldering, 10 seconds)	300°C

Electrical Characteristics (T_A = 25°C, V_{CC} = +5V to +15V, unless otherwise specified)

PARAMETER	CONDITIONS	LM556 MIN	LM556 TYP	LM556 MAX	LM556C MIN	LM556C TYP	LM556C MAX	UNITS
Supply Voltage		4.5		18	4.5		16	V
Supply Current	V_{CC} = 5V, R_L = ∞		3	5		3	6	mA
(Each Timer Section)	V_{CC} = 15V, R_L = ∞		10	11		10	14	mA
	(Low State) (Note 2)							
Timing Error, Monostable								
Initial Accuracy			0.5			0.75		%
Drift With Temperature	R_A, R_B = 1k to 100k, C = 0.1µF,		30			50		ppm/°C
	(Note 3)							
Accuracy Over Temperature			1.5			1.5		%
Drift with Supply			0.05			0.1		%/V
Timing Error, Astable								
Initial Accuracy			1.5			2.25		%
Drift With Temperature			90			150		ppm/°C
Accuracy Over Temperature			2.5			3.0		%
Drift With Supply			0.15			0.30		%/V
Trigger Voltage	V_{CC} = 15V	4.8	5	5.2	4.5	5	0.5	V
	V_{CC} = 5V	1.45	1.67	1.9	1.25	1.67	2.0	V
Trigger Current			0.1	0.5		0.2	1.0	µA
Reset Voltage	(Note 4)	0.4	0.5	1	0.4	0.5	1	V
Reset Current			0.1	0.4		0.1	0.6	mA
Threshold Current	(Note 5)		0.03	0.1		0.03	0.1	µA
Control Voltage Level And	V_{CC} = 15V	9.6	10	10.4	9	10	11	V
Threshold Voltage	V_{CC} = 5V	2.9	3.33	3.8	2.6	3.33	4	V
Pin 1, 13 Leakage Output High			1	100		1	100	nA
Pin 1, 13 Sat	(Note 6)							
Output Low	V_{CC} = 15V, I = 15 mA		150	240		180	300	mV
Output Low	V_{CC} = 4.5V, I = 4.5 mA		70	100		80	200	mV
Output Voltage Drop (Low)	V_{CC} = 15V							
	I_{SINK} = 10 mA		0.1	0.15		0.1	0.25	V
	I_{SINK} = 50 mA		0.4	0.5		0.4	0.75	V
	I_{SINK} = 100 mA		2	2.25		2	2.75	V
	I_{SINK} = 200 mA		2.5			2.5		V
	V_{CC} = 5V							
	I_{SINK} = 8 mA		0.1	0.25				V
	I_{SINK} = 5 mA					0.25	0.35	V
Output Voltage Drop (High)	I_{SOURCE} = 200 mA, V_{CC} = 15V		12.5			12.5		V
	I_{SOURCE} = 100 mA, V_{CC} = 15V	13	13.3		12.75	13.3		V
	V_{CC} = 5V	3	3.3		2.75	3.3		V
Rise Time of Output			100			100		ns
Fall Time of Output			100			100		ns
Matching Characteristics	(Note 7)							
Initial Timing Accuracy			0.05	0.2		0.1	2.0	%
Timing Drift With Temperature			±10			±10		ppm/°C
Drift With Supply Voltage			0.1	0.2		0.2	0.5	%/V

Note 1: For operating at elevated temperatures the device must be derated based on a +150°C maximum junction temperature and a thermal resistance of +150°C/W junction to ambient for both packages.

Note 2: Supply current when output high typically 1 mA less at V_{CC} = 5V.

Note 3: Tested at V_{CC} = 5V and V_{CC} = 15V.

Note 4: As reset voltage lowers, timing is inhibited and then the output goes low.

Note 5: This will determine the maximum value of R_A + R_B for 15V operation. The maximum total (R_A + R_B) is 20 MΩ.

Note 6: No protection against excessive pin 1, 13 current is necessary providing the package dissipation rating will not be exceeded.

Note 7: Matching characteristics refer to the difference between performance characteristics of each timer section.

National Semiconductor

LM565/LM565C Phase Locked Loop

General Description

The LM565 and LM565C are general purpose phase locked loops containing a stable, highly linear voltage controlled oscillator for low distortion FM demodulation, and a double balanced phase detector with good carrier suppression. The VCO frequency is set with an external resistor and capacitor, and a tuning range of 10:1 can be obtained with the same capacitor. The characteristics of the closed loop system—bandwidth, response speed, capture and pull in range—may be adjusted over a wide range with an external resistor and capacitor. The loop may be broken between the VCO and the phase detector for insertion of a digital frequency divider to obtain frequency multiplication.

The LM565H is specified for operation over the –55°C to +125°C military temperature range. The LM565CH and LM565CN are specified for operation over the 0°C to +70°C temperature range.

Features

■ 200 ppm/°C frequency stability of the VCO

■ Power supply range of ±5 to ±12 volts with 100 ppm/% typical
■ 0.2% linearity of demodulated output
■ Linear triangle wave with in phase zero crossings available
■ TTL and DTL compatible phase detector input and square wave output
■ Adjustable hold in range from ±1% to > ±60%.

Applications

■ Data and tape synchronization
■ Modems
■ FSK demodulation
■ FM demodulation
■ Frequency synthesizer
■ Tone decoding
■ Frequency multiplication and division
■ SCA demodulators
■ Telemetry receivers
■ Signal regeneration
■ Coherent demodulators.

Schematic and Connection Diagrams

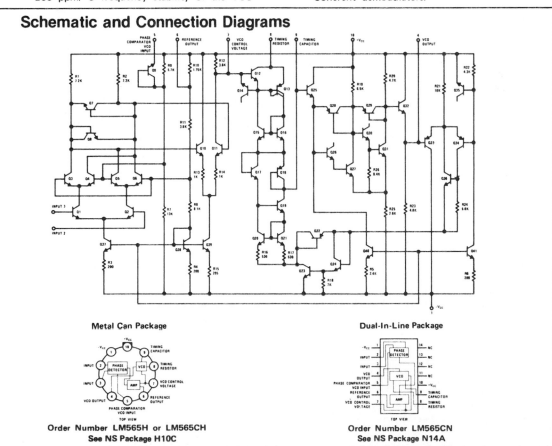

Metal Can Package

Order Number LM565H or LM565CH
See NS Package H10C

Dual-In-Line Package

Order Number LM565CN
See NS Package N14A

Absolute Maximum Ratings

Supply Voltage	±12V
Power Dissipation (Note 1)	300 mW
Differential Input Voltage	±1V
Operating Temperature Range LM565H	−55°C to +125°C
LM565CH, LM565CN	0°C to 70°C
Storage Temperature Range	−65°C to +150°C
Lead Temperature (Soldering, 10 sec)	300°C

Electrical Characteristics (AC Test Circuit, T_A = 25°C, V_C = ±6V)

PARAMETER	CONDITIONS	LM565			LM565C			UNITS
		MIN	TYP	MAX	MIN	TYP	MAX	
Power Supply Current			8.0	12.5		8.0	12.5	mA
Input Impedance (Pins 2, 3)	−4V < V_2, V_3 < 0V	7	10			5		kΩ
VCO Maximum Operating Frequency	C_o = 2.7 pF	300	500		250	500		kHz
Operating Frequency Temperature Coefficient			−100	300		−200	500	ppm/°C
Frequency Drift with Supply Voltage			0.01	0.1		0.05	0.2	%/V
Triangle Wave Output Voltage		2	2.4	3	2	2.4	3	V_{p-p}
Triangle Wave Output Linearity			0.2	0.75		0.5	1	%
Square Wave Output Level		4.7	5.4		4.7	5.4		V_{p-p}
Output Impedance (Pin 4)			5			5		kΩ
Square Wave Duty Cycle		45	50	55	40	50	60	%
Square Wave Rise Time			20	100		20		ns
Square Wave Fall Time			50	200		50		ns
Output Current Sink (Pin 4)		0.6	1		0.6	1		mA
VCO Sensitivity	f_o = 10 kHz	6400	6600	6800	6000	6600	7200	Hz/V
Demodulated Output Voltage (Pin 7)	±10% Frequency Deviation	250	300	350	200	300	400	mV_{pp}
Total Harmonic Distortion	±10% Frequency Deviation		0.2	0.75		0.2	1.5	%
Output Impedance (Pin 7)			3.5			3.5		kΩ
DC Level (Pin 7)		4.25	4.5	4.75	4.0	4.5	5.0	V
Output Offset Voltage $|V_7 − V_6|$			30	100		50	200	mV
Temperature Drift of $|V_7 − V_6|$			500			500		μV/°C
AM Rejection		30	40			40		dB
Phase Detector Sensitivity K_D		0.6	.68	0.9	0.55	.68	0.95	V/radian

Note 1: The maximum junction temperature of the LM565 is 150°C, while that of the LM565C and LM565CN is 100°C. For operation at elevated temperatures, devices in the TO-5 package must be derated based on a thermal resistance of 150°C/W junction to ambient or 45°C/W junction to case. Thermal resistance of the dual-in-line package is 100°C/W.

National Semiconductor

LM566/LM566C Voltage Controlled Oscillator

General Description

The LM566/LM566C are general purpose voltage controlled oscillators which may be used to generate square and triangular waves, the frequency of which is a very linear function of a control voltage. The frequency is also a function of an external resistor and capacitor.

The LM566 is specified for operation over the –55°C to +125°C military temperature range. The LM566C is specified for operation over the 0°C to +70°C temperature range.

Features

- Wide supply voltage range: 10 to 24 volts
- Very linear modulation characteristics
- High temperature stability
- Excellent supply voltage rejection
- 10 to 1 frequency range with fixed capacitor
- Frequency programmable by means of current, voltage, resistor or capacitor.

Applications

- FM modulation
- Signal generation
- Function generation
- Frequency shift keying
- Tone generation

Schematic and Connection Diagrams

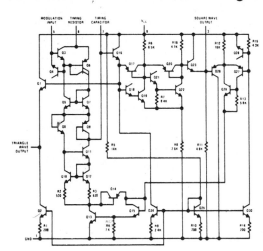

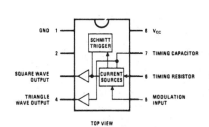

TOP VIEW

**Order Number LM566CN
See NS Package N08B**

Typical Application

**1 kHz and 10 kHz TTL Compatible
Voltage Controlled Oscillator**

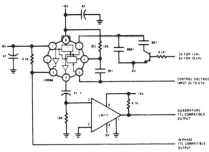

Applications Information

The LM566 may be operated from either a single supply as shown in this test circuit, or from a split (±) power supply. When operating from a split supply, the square wave output (pin 4) is TTL compatible (2 mA current sink) with the addition of a 4.7 kΩ resistor from pin 3 to ground.

A .001 μF capacitor is connected between pins 5 and 6 to prevent parasitic oscillations that may occur during VCO switching.

$$f_O = \frac{2(V^+ - V_5)}{R_1 C_1 V^+}$$

where

$2K < R_1 < 20K$

and V_5 is voltage between pin 5 and pin 1

Absolute Maximum Ratings

Power Supply Voltage	26V
Power Dissipation (Note 1)	300 mW
Operating Temperature Range LM566	-55°C to $+125^\circ$C
LM566C	0°C to 70°C
Lead Temperature (Soldering, 10 sec)	300°C

Electrical Characteristics V_{CC} = 12V, T_A = 25°C, AC Test Circuit

PARAMETER	CONDITIONS	LM566			LM566C			UNITS
		MIN	TYP	MAX	MIN	TYP	MAX	
Maximum Operating Frequency	R0 = 2k CO = 2.7 pF		1			1		MHz
Input Voltage Range Pin 5		3/4 V_{CC}		V_{CC}	3/4 V_{CC}		V_{CC}	
Average Temperature Coefficient of Operating Frequency			100			200		ppm/$^\circ$C
Supply Voltage Rejection	10-20V		0.1	1		0.1	2	%/V
Input Impedance Pin 5		0.5	1		0.5	1		MΩ
VCO Sensitivity	For Pin 5, From 8–10V, f_O = 10 kHz	6.4	6.6	6.8	6.0	6.6	7.2	kHz/V
FM Distortion	$\pm$10% Deviation		0.2	0.75		0.2	1.5	%
Maximum Sweep Rate		800	1		500	1		MHz
Sweep Range			10:1			10:1		
Output Impedance								
Pin 3			50			50		Ω
Pin 4			50			50		Ω
Square Wave Output Level	R_{L1} = 10k	5.0	5.4		5.0	5.4		Vp-p
Triangle Wave Output Level	R_{L2} = 10k	2.0	2.4		2.0	2.4		Vp-p
Square Wave Duty Cycle		45	50	55	40	50	60	%
Square Wave Rise Time			20			20		ns
Square Wave Fall Time			50			50		ns
Triangle Wave Linearity	+1V Segment at 1/2 V_{CC}		0.2	0.75		0.5	1	%

Note 1: The maximum junction temperature of the LM566 is 150°C, while that of the LM566C is 100°C. For operating at elevated junction temperatures, devices in the TO-5 package must be derated based on a thermal resistance of 150°C/W. The thermal resistance of the dual-in-line package is 100°C/W.

SIECOR/OPTICAL CABLE

SIECOR® FAT FIBER™
CABLE (144)

A SERVICE OF **SIECOR**

SIECOR 144 Fat Fiber cable is a lightweight, all-glass fiber optic cable designed for use in a variety of applications. It is optimized for distances of 200 to 2000 meters. The flame retardant SIECOR 144 Fat Fiber cable can be installed in industrial, process control, computer and wired office applications. Since this product has the strength, flexibility, and general handling characteristics found in miniature coaxial cables, it can be used in the same environments without EMI and ground loop problems.

Fat Fiber Cables

Reliable High Performance Systems Possible
- Low cost sources and detectors
- Stable transmission performance
- High numerical aperture, large core for improved coupling efficiency
- Straightforward transmitters and receivers
- Commercially available connectors

Excellent Environmental Performance
- Wide operating temperature range
- Flame retardant to meet safety standards (passes UL VW1 vertical flame test)

Convenient
- Easily strippable outer sheath and fiber buffer
- Easily installed — small, flexible, but rugged

Product Number 144

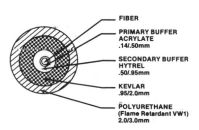

- FIBER
- PRIMARY BUFFER ACRYLATE .14/.50mm
- SECONDARY BUFFER HYTREL .50/.95mm
- KEVLAR .95/2.0mm
- POLYURETHANE (Flame Retardant VW1) 2.0/3.0mm

CABLE PROPERTIES

SIECOR/OPTICAL CABLE product number	**144**
Number of fibers	1
Maximum attenuation @ 850 nm	10 dB/km
Minimum bandwidth (-3dB) @ 850 nm	20 MHz·km

Installation and use

Operating temperature range, installed	− 20 to 85 °C
Maximum tensile load for installation	300 N*
Minimum bend radius for installation at 300 N	5 cm
Maximum tensile load, long term installation	50 N
Structural cable strength	1500 N

Additional data

Cable: Outside diameter, nominal	3.0 mm
Weight, nominal	7.5 kg/km
Fiber: Core diameter, nominal	100 µm
Cladding diameter, nominal	140 µm
Coated diameter, nominal	950 µm
Numerical aperture (100% short length), nominal	0.3

---

*One Newton = 0.102 kg_f = 0.225 lb_f

Duplex version (244) also available.

For additional information contact:
SIECOR/OPTICAL CABLE
Four Eighty Nine/Hickory, N.C. 28603
Telephone 704 322-3740
Telex: 800546
Cable: SIECOR

144P (5/81CP)

Printed in U.S.A.

FIBER OPTICS

PLASTIC CLAD SILICA OPTICAL FIBER   3/10/81

Plastic Clad Silica Optical Fiber is designed for medium to long distance and medium bandwidth data transmission as well as high power optical transmission applications. This fiber type exhibits the lowest radiation sensitivity reported. The fiber consists of a high purity silica core, plastic optical cladding and a protective jacket. The optical waveguide is comprised of a silica core and plastic cladding while the outer jacket provides environmental and mechanical protection. At -55°C this fiber will exhibit good low loss behavior.

TYPICAL SPECTRAL ATTENUATION

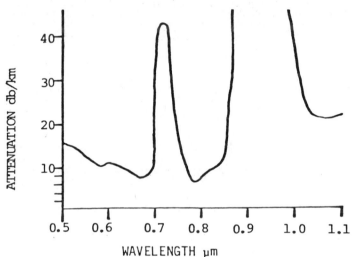

FIBER TYPE	EPC-BC200	EPC-BC400	EPC-BC600
ATTENUATION (790 nm)	6 db/km	6 db/km	6 db/km
NUMERICAL APERTURE	0.4	0.4	0.4
CORE INDEX OF REFRACTION	1.46	1.46	1.46
FIBER CORE DIAMETER (μm)	200	400	600
JACKET OUTER DIAMETER (μm)	500	750	1000
MINIMUM BEND RADIUS (cm)	0.8	1.6	2.8
-3 db INTERMODAL DISPERSION (MHZ-km)	20	15	10
FIBER TENSILE STRENGTH (0.5m GAUGE LENGTH) Nt/m^2	3.5×10^9 (500 KPSI)	3.5×10^9 (500 KPSI)	3.5×10^9 (500 KPSI)

Fiber Parameters Shown Are Nominal Values

EOTec
Corporation   200 Frontage Road • West Haven, Connecticut 06516 • (203) 934-7961

208

GATES

54/7400, LS00, S00

Quad Two-Input NAND Gate

TYPE	TYPICAL PROPAGATION DELAY	TYPICAL SUPPLY CURRENT (Total)
7400	9ns	8mA
74LS00	9.5ns	1.6mA
74S00	3ns	15mA

ORDERING CODE

PACKAGES	COMMERCIAL RANGES $V_{CC} = 5V \pm 5\%$; $T_A = 0°C$ to $+70°C$	MILITARY RANGES $V_{CC} = 5V \pm 10\%$; $T_A = -55°C$ to $+125°C$
Plastic DIP	N7400N • N74LS00N N74S00N	
Plastic SO	N74LS00D N74S00D	
Ceramic DIP		S5400F • S54LS00F S54S00F
Flatpack		S5400W • S54LS00W S54S00W
LLCC		S54LS00G

FUNCTION TABLE

INPUTS		OUTPUT
A	B	Y
L	L	H
L	H	H
H	L	H
H	H	L

H = HIGH voltage level
L = LOW voltage level

INPUT AND OUTPUT LOADING AND FAN-OUT TABLE

PINS	DESCRIPTION	54/74	54/74S	54/74LS
A, B	Inputs	1ul	1Sul	1LSul
Y	Output	10ul	10Sul	10LSul

NOTE
Where a 54/74 unit load (ul) is understood to be 40µA I_{IH} and −1.6mA I_{IL}, a 54/74S unit load (Sul) is 50µA I_{IH} and −2.0mA I_{IL}, and 54/74LS unit load (LSul) is 20µA I_{IH} and −0.4mA I_{IL}.

PIN CONFIGURATION

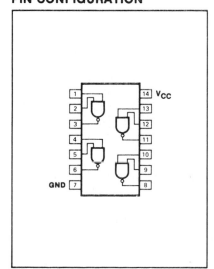

LOGIC SYMBOL

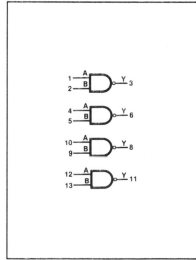

LOGIC SYMBOL (IEEE/IEC)

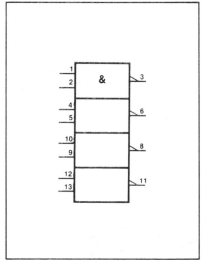

Signetics

GATES

DC ELECTRICAL CHARACTERISTICS (Over recommended operating free-air temperature range unless otherwise noted.)

PARAMETER		TEST CONDITIONS[1]		54/7400			54/74LS00			54/74S00			UNIT
				Min	Typ[2]	Max	Min	Typ[2]	Max	Min	Typ[2]	Max	
V_{OH}	HIGH-level output voltage	$V_{CC} = MIN$, $V_{IH} = MIN$, $V_{IL} = MAX$, $I_{OH} = MAX$	Mil	2.4	3.4		2.5	3.4		2.5	3.4		V
			Com'l	2.4	3.4		2.7	3.4		2.7	3.4		V
V_{OL}	LOW-level output voltage	$V_{CC} = MIN$, $V_{IH} = MIN$ $I_{OL} = MAX$	Mil		0.2	0.4		0.25	0.4			0.5[4]	V
			Com'l		0.2	0.4		0.35	0.5			0.5	V
		$I_{OL} = 4mA$	74LS					0.25	0.4				V
V_{IK}	Input clamp voltage	$V_{CC} = MIN$, $I_I = I_{IK}$				− 1.5			− 1.5			− 1.2	V
I_I	Input current at maximum input voltage	$V_{CC} = MAX$	$V_I = 5.5V$			1.0						1.0	mA
			$V_I = 7.0V$						0.1				mA
I_{IH}	HIGH-level input current	$V_{CC} = MAX$	$V_I = 2.4V$			40							μA
			$V_I = 2.7V$						20			50	μA
I_{IL}	LOW-level input current	$V_{CC} = MAX$	$V_I = 0.4V$			− 1.6			− 0.4				mA
			$V_I = 0.5V$									− 2.0	mA
I_{OS}	Short-circuit output current[3]	$V_{CC} = MAX$	Mil	− 20		− 55	− 20		− 100	− 40		− 100	mA
			Com'l	− 18		− 55	− 20		− 100	− 40		− 100	mA
I_{CC}	Supply current (total)	$V_{CC} = MAX$	I_{CCH} Outputs HIGH		4	8		0.8	1.6		10	16	mA
			I_{CCL} Outputs LOW		12	22		2.4	4.4		20	36	mA

NOTES
1. For conditions shown as MIN or MAX, use the appropriate value specified under recommended operating conditions for the applicable type.
2. All typical values are at $V_{CC} = 5V$, $T_A = 25°C$.
3. I_{OS} is tested with $V_{OUT} = + 0.5V$ and $V_{CC} = V_{CC}$ MAX + 0.5V. Not more than one output should be shorted at a time and duration of the short circuit should not exceed one second.
4. $V_{OL} = + 0.45V$ MAX for 54S at $T_A = + 125°C$ only.

AC WAVEFORM

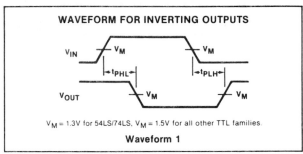

WAVEFORM FOR INVERTING OUTPUTS

$V_M = 1.3V$ for 54LS/74LS, $V_M = 1.5V$ for all other TTL families.

Waveform 1

AC CHARACTERISTICS $T_A = 25°C$, $V_{CC} = 5.0V$

PARAMETER		TEST CONDITIONS	54/74 $C_L = 15pF$, $R_L = 400Ω$		54/74LS $C_L = 15pF$, $R_L = 2kΩ$		54/74S $C_L = 15pF$, $R_L = 280Ω$		UNIT
			Min	Max	Min	Max	Min	Max	
t_{PLH} t_{PHL}	Propagation delay	Waveform 1		22 15		15 15		4.5 5.0	ns

INVERTERS

54/7404, LS04, S04

Hex Inverter

TYPE	TYPICAL PROPAGATION DELAY	TYPICAL SUPPLY CURRENT (Total)
7404	10ns	12mA
74LS04	9.5ns	2.4mA
74S04	3ns	22mA

ORDERING CODE

PACKAGES	COMMERCIAL RANGES $V_{CC} = 5V \pm 5\%$; $T_A = 0°C$ to $+70°C$	MILITARY RANGES $V_{CC} = 5V \pm 10\%$; $T_A = -55°C$ to $+125°C$
Plastic DIP	N7404N • N74LS04N N74S04N	
Plastic SO	N74LS04D • N74S04D	
Ceramic DIP		S5404F • S54LS04F S54S04F
Flatpack		S5404W • S54LS04W S54S04W
LLCC		S54LS04G

FUNCTION TABLE

INPUT	OUTPUT
A	Y
L	H
H	L

H = HIGH voltage level
L = LOW voltage level

INPUT AND OUTPUT LOADING AND FAN-OUT TABLE

PINS	DESCRIPTION	54/74	54/74S	54/74LS
A	Input	1ul	1Sul	1LSul
Y	Output	10ul	10Sul	10LSul

NOTE
Where a 54/74 unit load (ul) is understood to be 40µA I_{IH} and -1.6mA I_{IL}, a 54/74S unit load (Sul) is 50µA I_{IH} and -2.0mA I_{IL}, and 54/74LS unit load (LSul) is 20µA I_{IH} and -0.4mA I_{IL}.

PIN CONFIGURATION

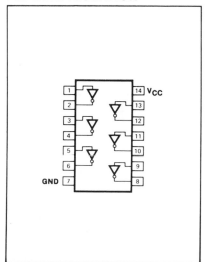

LOGIC SYMBOL

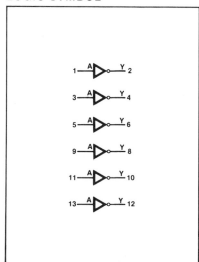

LOGIC SYMBOL (IEEE/IEC)

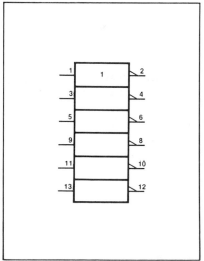

INVERTERS

ABSOLUTE MAXIMUM RATINGS (Over operating free-air temperature range unless otherwise noted.)

	PARAMETER	54	54LS	54S	74	74LS	74S	UNIT
V_{CC}	Supply voltage	7.0	7.0	7.0	7.0	7.0	7.0	V
V_{IN}	Input voltage	−0.5 to +5.5	−0.5 to +7.0	−0.5 to +5.5	−0.5 to +5.5	−0.5 to +7.0	−0.5 to +5.5	V
I_{IN}	Input current	−30 to +5	−30 to +1	−30 to +5	−30 to +5	−30 to +1	−30 to +5	mA
V_{OUT}	Voltage applied to output in HIGH output state	−0.5 to +V_{CC}	−0.5 to +V_{CC}	−0.5 to +V_{CC}	−0.5 to +V_{CC}	−0.5 to +V_{CC}	−0.5 to +V_{CC}	V
T_A	Operating free-air temperature range	−55 to +125			0 to 70			°C

RECOMMENDED OPERATING CONDITIONS

	PARAMETER		54/74 Min	54/74 Nom	54/74 Max	54/74LS Min	54/74LS Nom	54/74LS Max	54/74S Min	54/74S Nom	54/74S Max	UNIT
V_{CC}	Supply voltage	Mil	4.5	5.0	5.5	4.5	5.0	5.5	4.5	5.0	5.5	V
		Com'l	4.75	5.0	5.25	4.75	5.0	5.25	4.75	5.0	5.25	V
V_{IH}	HIGH-level input voltage		2.0			2.0			2.0			V
V_{IL}	LOW-level input voltage	Mil			+0.8			+0.7			+0.8	V
		Com'l			+0.8			+0.8			+0.8	V
I_{IK}	Input clamp current				−12			−18			−18	mA
I_{OH}	HIGH-level output current				−400			−400			−1000	μA
I_{OL}	LOW-level output current	Mil			16			4			20	mA
		Com'l			16			8			20	mA
T_A	Operating free-air temperature	Mil	−55		+125	−55		+125	−55		+125	°C
		Com'l	0		70	0		70	0		70	°C

NOTE
V_{IL} = +0.7V MAX for 54S at T_A = +125°C only.

TEST CIRCUITS AND WAVEFORMS

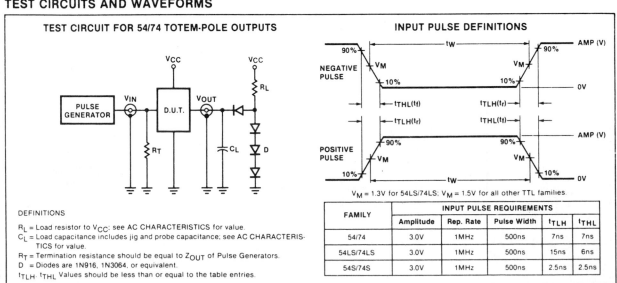

TEST CIRCUIT FOR 54/74 TOTEM-POLE OUTPUTS

INPUT PULSE DEFINITIONS

V_M = 1.3V for 54LS/74LS; V_M = 1.5V for all other TTL families.

DEFINITIONS

R_L = Load resistor to V_{CC}: see AC CHARACTERISTICS for value.
C_L = Load capacitance includes jig and probe capacitance; see AC CHARACTERISTICS for value.
R_T = Termination resistance should be equal to Z_{OUT} of Pulse Generators.
D = Diodes are 1N916, 1N3064, or equivalent.
t_{TLH}, t_{THL} Values should be less than or equal to the table entries.

FAMILY	INPUT PULSE REQUIREMENTS				
	Amplitude	Rep. Rate	Pulse Width	t_{TLH}	t_{THL}
54/74	3.0V	1MHz	500ns	7ns	7ns
54LS/74LS	3.0V	1MHz	500ns	15ns	6ns
54S/74S	3.0V	1MHz	500ns	2.5ns	2.5ns

Signetics

COUNTERS

Decade Counter

DESCRIPTION

The '90 is a 4-bit, ripple-type Decade Counter. The device consists of four master-slave flip-flops internally connected to provide a divide-by-two section and a divide-by-five section. Each section has a separate Clock input to initiate state changes of the counter on the HIGH-to-LOW clock transition. State changes of the Q outputs do not occur simultaneously because of internal ripple delays. Therefore, decoded output signals are subject to decoding spikes and should not be used for clocks or strobes.

A gated AND asynchronous Master Reset ($MR_1 \cdot MR_2$) is provided which overrides both clocks and resets (clears) all the flip-flops. Also provided is a gated AND asynchronous Master Set ($MS_1 \cdot MS_2$) which overrides the clocks and the MR inputs, setting the outputs to nine (HLLH).

Since the output from the divide-by-two section is not internally connected to the succeeding stages, the device may be operated in various counting modes. In a BCD (8421) counter the $\overline{CP}_1$ input must be externally connected to the Q_0 output. The $\overline{CP}_0$ input receives the incoming count producing a BCD count sequence. In a symmetrical Bi-quinary divide-by-ten counter the Q_3 output must be connected externally to the $\overline{CP}_0$ input. The input count is then applied to the CP_1 input and a divide-by-ten square wave is obtained at output Q_0. To operate as a divide-by-two and a divide-by-five counter no external interconnections are required. The first flip-flop is used as a binary element for the divide-by-two function ($\overline{CP}_0$ as the input and Q_0 as the output). The $\overline{CP}_1$ input is used to obtain a divide-by-five operation at the Q_3 output.

TYPE	TYPICAL f_{MAX}	TYPICAL SUPPLY CURRENT
7490	30MHz	30mA
74LS90	42MHz	9mA

ORDERING CODE

PACKAGES	COMMERCIAL RANGES $V_{CC} = 5V \pm 5\%$; $T_A = 0°C$ to $+70°C$	MILITARY RANGES $V_{CC} = 5V \pm 10\%$; $T_A = -55°C$ to $+125°C$
Plastic DIP	N7490N • N74LS90N	
Ceramic DIP		S54LS90F
Flatpack		S54LS90W

INPUT AND OUTPUT LOADING AND FAN-OUT TABLE

PINS	DESCRIPTION	54/74	54/74LS
$\overline{CP}_0$	Input	2ul	6LSul
$\overline{CP}_1$	Input	4ul	8LSul
MR, MS	Inputs	1ul	1LSul
Q_0-Q_3	Outputs	10ul	10LSul

NOTE
Where a 54/74 unit load (ul) is understood to be 40µA I_{IH} and −1.6mA I_{IL}, and a 54/74LS unit load (LSul) is 20µA I_{IH} and −0.4mA I_{IL}.

PIN CONFIGURATION

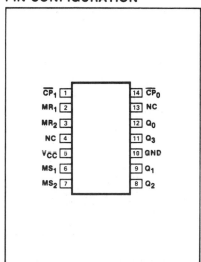

LOGIC SYMBOL

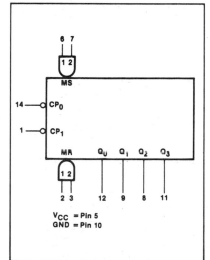

LOGIC SYMBOL (IEEE/IEC)

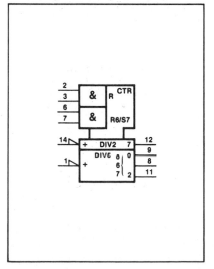

Signetics